BIRDS IN TOWN & VILLAGE

W. H. HUDSON

ZINC READ

Published by
This edition copyright © 2024 by Zinc Read
Design and setting By Zinc Read
Email- zincread@gmail.com
ISBN-978-93-6277-005-9

CONTENTS

PREFACE

THIS book is more than a mere reprint of *Birds in a Village* first published in 1893. That was my first book about bird life, with some impressions of rural scenes, in England; and, as is often the case with a first book, its author has continued to cherish a certain affection for it. On this account it pleased me when its turn came to be reissued, since this gave me the opportunity of mending some faults in the portions retained and of throwing out a good deal of matter which appeared to me not worth keeping.

The first portion, "Birds in a Village," has been mostly rewritten with some fresh matter added, mainly later observations and incidents introduced in illustration of the various subjects discussed. For the concluding portion of the old book, which has been discarded, I have substituted entirely new matter-the part entitled "Birds in a Cornish Village."

Between these two long parts there are five shorter essays which I have retained with little alteration, and these in one or two instances are consequently out of date, especially in what was said with bitterness in the essay on "Exotic Birds for Britain" anent the feather-wearing fashion and of the London trade in dead birds and the refusal of women at that time to help us in trying to save the beautiful wild bird life of this country and of the world generally from extermination. Happily, the last twenty years of the life and work of the Royal Society for the Protection of Birds have changed all that, and it would not now be too much to say that all right-thinking persons in this country, men and women, are anxious to see the end of this iniquitous traffic.

W. H. H.

September, 1919.

BIRDS IN TOWN & VILLAGE

I

ABOUT the middle of last May, after a rough and cold period, there came a spell of brilliant weather, reviving in me the old spring feeling, the passion for wild nature, the desire for the companionship of birds; and I betook myself to St. James's Park for the sake of such satisfaction as may be had from watching and feeding the fowls, wild and semi-wild, found gathered at that favored spot.

I was glad to observe a couple of those new colonists of the ornamental water, the dabchicks, and to renew my acquaintance with the familiar, long-established moorhens. One of them was engaged in building its nest in an elm-tree grow-

ing at the water's edge. I saw it make two journeys with large wisps of dry grass in its beak, running up the rough, slanting trunk to a height of sixteen to seventeen feet, and disappearing within the "brushwood sheaf" that springs from the bole at that distance from the roots. The woodpigeons were much more numerous, also more eager to be fed. They seemed to understand very quickly that my bread and grain was for them and not the sparrows; but although they stationed themselves close to me, the little robbers we were jointly trying to outwit managed to get some pieces of bread by flying up and catching them before they touched the sward. This little comedy over, I visited the water-fowl, ducks of many kinds, sheldrakes, geese from many lands, swans black, and swans white. To see birds in prison during the spring mood of which I have spoken is not only no satisfaction but a positive pain; here -- albeit without that large liberty that nature gives, they are free in a measure; and swimming and diving or dozing in the sunshine, with the blue sky above them, they are perhaps unconscious of any restraint. Walking along the margin I noticed three children some yards ahead

of me; two were quite small, but the third, in whose charge the others were, was a robust-looking girl, aged about ten or eleven years. From their dress and appearance I took them to be the children of a respectable artisan or small tradesman; but what chiefly attracted my attention was the very great pleasure the elder girl appeared to take in the birds. She had come well provided with stale bread to feed them, and after giving moderately of her store to the woodpigeons and sparrows, she went on to the others, native and exotic, that were disporting themselves in the water, or sunning themselves on the green bank. She did not cast her bread on the water in the manner usual with visitors, but was anxious to feed all the different species, or as many as she could attract to her, and appeared satisfied when any one individual of a particular kind got a fragment of her bread. Meanwhile she talked eagerly to the little ones, calling their attention to the different birds. Drawing near, I also became an interested listener; and then, in answer to my questions, she began telling me what all these strange fowls were. "This," she said, glad to give information, "is the Canadian goose, and

4 BIRDS IN TOWN AND VILLAGE

there is the Egyptian goose; and here is the king-duck coming towards us; and do you see that large, beautiful bird standing by itself, that will not come to be fed? That is the golden duck. But that is not its real name; I don't know them all, and so I name some for myself. I call that one the golden duck because in the sun its feathers sometimes shine like gold." It was a rare pleasure to listen to her, and seeing what sort of a girl she was, and how much in love with her subject, I in my turn told her a great deal about the birds before us, also of other birds she had never seen nor heard of, in other and distant lands that have a nobler bird life than ours; and after she had listened eagerly for some minutes, and had then been silent a little while, she all at once pressed her two hands together, and exclaimed rapturously, "Oh, I do so love the birds!"

I replied that that was not strange, since it is impossible for us not to love whatever is lovely, and of all living things birds were made most beautiful.

Then I walked away, but could not forget the words she had exclaimed, her whole appearance,

BIRDS IN A VILLAGE 5

the face flushed with color, the eloquent brown eyes sparkling, the pressed palms, the sudden spontaneous passion of delight and desire in her tone. The picture was in my mind all that day, and lived through the next, and so wrought on me that I could not longer keep away from the birds, which I, too, loved; for now all at once it seemed to me that life was not life without them; that I was grown sick, and all my senses dim; that only the wished sight of wild birds could medicine my vision; that only by drenching it in their wild melody could my tired brain recover its lost vigour.

II

AFTER wandering somewhat aimlessly about the country for a couple of days, I stumbled by chance on just such a spot as I had been wishing to find--a rustic village not too far away. It was not more than twenty-five minutes' walk from a small station, less than one hour by rail from London.

The way to the village was through cornfields, bordered by hedges and rows of majestic elms. Beyond it, but quite near, there was a wood, principally of beech, over a mile in length, with a public path running through it. On the right hand, ten minutes' walk from the village, there was a long green hill, the ascent to which was gentle; but on the further side it sloped abruptly down to the Thames.

On the left hand there was another hill, with cottages and orchards, with small fields interspersed on the slope and summit, so that the

middle part, where I lodged, was in a pretty deep hollow. There was no sound of traffic there, and few farmers' carts came that way, as it was well away from the roads, and the deep, narrow, winding lanes were exceedingly rough, like the stony beds of dried-up streams.

In the deepest part of the coombe, in the middle of the village, there was a well where the cottagers drew their water; and in the summer evenings the youths and maidens came there, with or without jugs and buckets, to indulge in conversation, which was mostly of the rustic, bantering kind, mixed with a good deal of loud laughter. Close by was the inn, where the men sat on benches in the tap-room in grave discourse over their pipes and beer.

Wishing to make their acquaintance, I went in and sat down among them, and found them a little shy--not to say stand-offish, at first. Rustics are often suspicious of the stranger within their gates; but after paying for beer all round, the frost melted and we were soon deep in talk about the wild life of the place; always a safe and pleasant subject in a village. One rough-looking, brown-faced man, with iron-grey hair,

---

became a sort of spokesman for the company, and replied to most of my questions.

"And what about badgers?" I asked. "In such a rough-looking spot with woods and all, it strikes me as just the sort of place where one would find that animal."

A long dead silence followed. I caught the eye of the man nearest me and repeated the question, "Are there no badgers here?" His eyes fell, then he exchanged glances with some of the others, all very serious; and at length my man, addressing the person who had acted as spokesman before, said, "Perhaps you'll tell the gentleman if there are any badgers here."

At that the rough man looked at me very sharply, and answered stiffly, "Not as I know of."

A few weeks later, at a small town in the neighbourhood, I got into conversation with a hotel keeper, an intelligent man, who gave me a good deal of information about the country. He asked me where I was staying, and, on my telling him, said "Ah, I know it well--that village in a hole; and a very nasty hole to get in, too--at any rate it was so, formerly. They are getting a bit civilized now, but I remember the, time

---

when a stranger couldn't show himself in the place without being jeered at and insulted. Yes, they were a rough lot down in that hole--the Badgers, they were called, and that's what they are called still."

The pity of it was that I didn't know this before I went among them! But it was not remembered against me that I had wounded their susceptibilities; they soon found that I was nothing but a harmless field naturalist, and I had friendly relations with many of them.

At the extremity of the straggling village was the beginning of an extensive common, where it was always possible to spend an hour or two without seeing a human creature. A few sheep grazed and browsed there, roaming about in twos and threes and half-dozens, tearing their fleeces for the benefit of nest-building birds, in the great tangled masses of mingled furze and bramble and briar. Birds were abundant there--all those kinds that love the common's openness, and the rough, thorny vegetation that flourishes on it. But the village--or rather, the large open space occupied by it, formed the headquarters and centre of a paradise of birds (as I soon began to think

it), for the cottages and houses were widely separated, the meanest having a garden and some trees, and in most cases there was an old orchard of apple, cherry, and walnut trees to each habitation, and out of this mass of greenery, which hid the houses and made the place look more like a wood than a village, towered the great elms in rows, and in groups.

On first approaching the place I heard, mingled with many other voices, that of the nightingale; and as it was for the medicine of its pure, fresh melody that I particularly craved, I was glad to find a lodging in one of the cottages, and to remain there for several weeks.

The small care which the nightingale took to live up to his reputation in this place surprised me a little. Here he could always be heard in the daytime--not one bird, but a dozen--in different parts of the village; but he sang not at night. This I set down to the fact that the nights were dark and the weather unsettled. But later, when the weather grew warmer, and there were brilliant moonlight nights, he was still a silent bird except by day.

I was also a little surprised at his tameness.

NIGHTINGALE.
"the medicine of its pure, fresh melody"

BIRDS IN A VILLAGE 11

On first coming to the village, when I ran after every nightingale I heard, to get as near him as possible, I was occasionally led by the sound to a cottage, and in some instances I found the singer perched within three or four yards of an open window or door. At my own cottage, when the woman who waited on me shook the breakfast cloth at the front door, the bird that came to pick up the crumbs was the nightingale--not the robin. When by chance he met a sparrow there, he attacked and chased it away. It was a feast of nightingales. An elderly woman of the village explained to me that the nightingales and other small birds were common and tame in the village, because no person disturbed them. I smile now when recording the good old dame's words.

On my second day at the village it happened to be raining--a warm, mizzling rain without wind--ind the nightingales were as vocal as in fine bright weather. I heard one in a narrow lane, and went towards it, treading softly, in order not to scare it away, until I got within eight or ten yards of it, as it sat on a dead projecting twig. This was a twig of a low thorn

12 BIRDS IN TOWN AND VILLAGE

tree growing up from the hedge, projecting through the foliage, and the bird, perched near its end, sat only about five feet above the bare ground of the lane. Now, I owe my best thanks to this individual nightingale, for sharply calling to my mind a common pestilent delusion, which I have always hated, but had never yet raised my voice against--namely, that all wild creatures exist in constant fear of an attack from the numberless subtle or powerful enemies that are always waiting and watching for an opportunity to spring upon and destroy them. The truth is, that although their enemies be legion, and that every day, and even several times on each day, they may be threatened with destruction, they are absolutely free from apprehension, except when in the immediate presence of danger. Suspicious they may be at times, and the suspicion may cause them to remove themselves to a greater distance from the object that excites it; but the emotion is so slight, the action so almost automatic, that the singing bird will fly to another bush a dozen yards away, and at once resume his interrupted song. Again, a bird will see the deadliest enemy of its kind, and unless it be so

BIRDS IN A VILLAGE 13

close as to actually threaten his life, he will regard it with the greatest indifference or will only be moved to anger at its presence. Here was this nightingale singing in the rain, seeing but not heeding me; while beneath the hedge, almost directly under the twig it sat on, a black cat was watching it with luminous yellow eyes. I did not see the cat at first, but have no doubt that the nightingale had seen and knew that it was there. High up on the tops of the thorn, a couple of sparrows were silently perched. Perhaps, like myself, they had come there to listen. After I had been standing motionless, drinking in that dulcet music for at least five minutes, one of the two sparrows dropped from the perch straight down, and alighting on the bare wet ground directly under the nightingale, began busily pecking at something eatable it had discovered. No sooner had he begun pecking than out leaped the concealed cat on to him. The sparrow fluttered wildly up from beneath or between the claws, and escaped, as if by a miracle. The cat raised itself up, glared round, and, catching sight of me close by, sprang back into the hedge and was gone. But all this time

---

the exposed nightingale, perched only five feet above the spot where the attack had been made and the sparrow had so nearly lost his life, had continued singing; and he sang on for some minutes after. I suppose that he had seen the cat before, and knew instinctively that he was beyond its reach; that it was a terrestrial, not an aerial enemy, and so feared it not at all; and he would, perhaps, have continued singing if the sparrow had been caught and instantly killed.

Quite early in June I began to feel just a little cross with the nightingales, for they almost ceased singing; and considering that the spring had been a backward one, it seemed to me that their silence was coming too soon. I was not sufficiently regardful of the fact that their lays are solitary, as the poet has said; that they ask for no witness of their song, nor thirst for human praise. They were all nesting now. But if I heard them less, I saw much more of them, especially of one individual, the male bird of a couple that had made their nest in a hedge a stone's throw from the cottage. A favourite morning perch of this bird was on a small wooden gate four or five yards away from my window. It was an open, sunny spot, where his restless, bright eyes could sweep the lane, up and down; and he could there also give vent to his superfluous energy by lording it over a few sparrows and other small birds that visited the spot. I greatly admired the fine, alert figure of the pugnacious little creature, as he perched there so close to me, and so fearless. His striking resemblance to the robin in form, size, and in his motions, made his extreme familiarity seem only natural. The robin is greatly distinguished in a sober-plumaged company by the vivid tint on his breast. He is like the autumn leaf that catches a ray of sunlight on its surface, and shines conspicuously among russet leaves. But the clear brown of the nightingale is beautiful, too.

This same nightingale was keeping a little surprise in store for me. Although he took no notice of me sitting at the open window, whenever I went thirty or forty yards from the gate along the narrow lane that faced it, my presence troubled him and his mate only too much. They would flit round my head, emitting the two strongly contrasted sounds with which they express solicitude--the clear, thin, plaintive, or

wailing note, and the low, jarring sound--an alternate lamenting and girding. One day when I approached the nest, they displayed more anxiety than usual, fluttering close to me, wailing and croaking more vehemently than ever, when all at once the male, at the height of his excitement, burst into singing. Half a dozen notes were uttered rapidly, with great strength, then a small complaining cry again, and at intervals, a fresh burst of melody. I have remarked the same thing in other singing birds, species in which the harsh grating or piercing sounds that properly express violent emotions of a painful kind, have been nearly or quite lost. In the nightingale, this part of the bird's language has lost its original character, and has dwindled to something very small. Solicitude, fear, anger, are expressed with sounds that are mere lispings compared with those emitted by the bird when singing. It is worthy of remark that some of the most highly developed melodists--and I am now thinking of the mocking-birds--never, in -moments of extreme agitation, fall into this confusion and use singing notes that express agreeable emotions, to express such as are painful. But in the

---

BIRDS IN A VILLAGE 17

mocking-bird the primitive harsh and grating cries have not been lost nor softened to sounds hardly to be distinguished from those that are

emitted by way of song.

III

BY this time all the birds were breeding, some already breeding a second time. And now I began to suspect that they were not quite so undisturbed as the old dame had led me to believe; that they had not found a paradise in the village after all. One morning, as I moved softly along the hedge in my nightingale's lane, all at once I heard, in the old grassy orchard, to which it formed a boundary, swishing sounds of scuttling feet and half-suppressed exclamations of alarm; then a crushing through the hedge, and out, almost at my feet, rushed and leaped and tumbled half-a-dozen urchins, who had suddenly been frightened from a bird-nesting raid. Clothes torn, hands and faces scratched with thorns, hat-less, their tow-coloured hair all disordered or standing up like a white crest above their brown faces, rounded eyes staring--what an extraordinarily wild appearance they had! I was back

---

in very old times, in the Britain of a thousand years before the coming of the Romans, and these were her young barbarians, learning their life's business in little things.

No, the birds of the village were not undisturbed while breeding; but happily the young savages never found my nightingale's nest. One day the bird came to the gate as usual, and was more alert and pugnacious than ever; and no wonder, for his mate came too, and with them four young birds. For a week they were about the cottage every day, when they dispersed, and one beautiful bright morning the male bird, in his old place near my window, attempted to sing, beginning with that rich, melodious throbbing, which is usually called "*jugging,*" and following with half-a-dozen beautiful notes. That was all. It was July, and I heard no more music from him or from any other of his kind.

*

I have perhaps written at too great length of this bird. The nightingale was after all only one of the fifty-nine species I succeeded in identifying during my sojourn at the village. There were more. I heard the calls and cries of others in the wood and various places, but refused, except in the case of the too elusive crake, to set down any in my list that I did not see. It was not my ambition to make a long list. My greatest desire was to see well those that interested me most. But those who go forth, as I did, to look for birds that are a sight for sore eyes, must meet with many a disappointment. In all those fruit and shade trees that covered the village with a cloud of verdure, and in the neighbouring woods, not once did I catch a glimpse of the green woodpecker, a beautiful conspicuous bird, supposed to be increasing in many places in England. Its absence from so promising a locality seemed strange. Another species, also said to be increasing in the country--the turtledove, was extremely abundant. In the tall beech woods its low, monotonous crooning note was heard all day long from all sides. In shady places, where the loud, shrill bird-voices are few, one prefers this sound to the set song of the woodpigeon, being more continuous and soothing, and of the nature of a lullaby. It sometimes reminded me of the low monotone I

have heard from a Patagonian mother when singing her "swart papoose" to sleep. Still, I would gladly have spared many of these woodland crooners for the sake of one magpie--that bird of fine feathers and a bright mind, which I had not looked on for a whole year, and now hoped to see again. But he was not there; and after I had looked for myself, some of the natives assured me that no magpie had been seen for years in that wood.

For a time I feared that I was to be just as unlucky with regard to the jay, seeing that the owner of the extensive beech woods adjoining the village permitted his keeper to kill the most interesting birds in it--kestrels and sparrowhawks, owls, jays, and magpies. He was a new man, comparatively, in the place, and wanted to increase his preserves, but to do this it was necessary first to exclude the villagers--the Badgers, who were no doubt partial to pheasants' eggs. Now, to close an ancient right-of-way is a ticklish business, and this was an important one, seeing that the village women did their Saturday marketing in the town beyond the wood and river, and with the path closed they would have two miles

---

further to walk. The new lord wisely took this into consideration, and set himself to win the goodwill of the people before attempting any strong measures. He walked in the lanes and was affable to the cottage women and nice to the children, and by and bye he exclaimed, "What! No institute! no hall, or any place where you can meet and spend the long winter evenings? Well, I'll soon see to that." And soon, to their delight, they had a nice building reared on a piece of land which he bought for the purpose, furnished with tables, chairs, bagatelle boards, and all accessories; and he also supplied them with newspapers and magazines. He was immensely popular, but appeared to think little of what he had done. When they expressed their gratitude to him he would move his hand, and answer, "Oh, I'm going to do a great deal more than that for you!"

A few months went by, then he caused a notice to be put up about the neighbourhood that the path through the wood was going to be closed "by order." No one took any notice, and a few weeks later his workmen appeared on the scene and erected a huge oakwood barrier across the

---

path; also a notice on a board that the wood was strictly private and trespassers would be prosecuted. The villagers met in force at the institute and the inn that evening, and after discussing the matter over their ale, they armed themselves with axes and went in a body and demolished the barrier.

The owner was disgusted, but took no action. "This," he said, "is their gratitude"; and from that day he ceased to subscribe to the local charities or take his walks in the village. He had given the institute, and so could not pull it down nor prevent them from using it.

It was refreshing to hear that the Badgers had shown a proper spirit in the matter, and I was grateful to them for having kept the right-of-way, as on most days I spent several hours in the beautiful woods.

To return to the jay. In spite of the keeper's persecution, I knew that he was there; every morning when I got up to look out of the window between four and five o'clock, I heard from some quarter of the village that curious subdued, but far-reaching, scolding note he is accustomed to utter when his suspicions have been aroused.

---

24 BIRDS IN TOWN AND VILLAGE

That was the jay's custom--to come from the woods before even the earliest risers were up, and forage in the village. By and bye I discovered that, by lying motionless for an hour or so on the dry moss in the wood, he would at length grow so bold as to allow himself to be seen, but high up among the topmost branches. Then, by means of my binocular, I had the wild thing on my thumb, so to speak, exhibiting himself to me, inquisitive, perplexed, suspicious, enraged by turns, as he flirted wings and tail, lifted and lowered his crest, glancing down with bright, wild eyes. What a beautiful hypocrisy and delightful power this is which enables us, sitting or lying motionless, feigning sleep perhaps, thus to fool this wild, elusive creature, and bring all its cunning to naught! He is so much smaller and keener-sighted, able to fly, to perch far up above me, to shift his position every minute or two, masking his small figure with this or that tuft of leaves, while still keeping his eyes on me--in spite of it all to have him so close, and without moving or taking any trouble, to see him so much better than he can see me! But this is a legitimate trickery of science, so innocent that we

---

JAY.
. inquisitive, perplexed, suspicious, enraged by turns."

---

can laugh at our dupe when we practise it; nor do we afterwards despise our superior cunning and feel ashamed, as when we slaughter wild birds with far-reaching shot, which they cannot escape.

*

All these corvine birds, which the gamekeeper pursues so relentlessly, albeit they were before him, killing when they killed to better purpose; and, let us hope, will exist after him--all these must greatly surpass other kinds in sagacity to have escaped extermination. In the present condition of things, the jay is perhaps the best off, on account of his smaller size and less conspicuous colouring; but whether more cunning than the crow or magpie or not, in perpetual alertness and restless energy or intensity of life, he is without an equal among British birds. And this quality forms his chief attraction; it is more to the mind than his lifted crest and bright eyes, his fine vinaceous brown and the patch of sky-blue on his wings. One would miss him greatly from the woods; some of the melody may well

---

be spared for the sake of the sudden, brain-piercing, rasping, rending scream with which he startles us in our solitary forest walks.

It is this extreme liveliness of the jay which makes it more distressing to the mind to see it pent in a cage than other birds of its family, such as the magpie; just as it is more distressing to see a skylark than a finch in prison, because the lark has an irresistible impulse to rise when his singing fit is on. Sing he must, in or out of prison, yet there can be little joy in the performance when the bird is incessantly teased with the unsatisfied desire to mount and pour out his music at heaven's gate.

Out of the cages, jays make charming and beautiful pets, and some who have kept them have assured me that they are not mischievous birds. The late Mark Melford one time when I visited him, had two jays, handsome birds, in bright, glossy plumage, always free to roam where they liked, indoors or out. We were sitting talking in his garden when one of the jays came flying to us and perched on a wooden ledge a few feet from and above our heads, and after sitting quietly for a little while he suddenly made a dash

---

at my head, just brushing it with his wings, then returned to his perch. At intervals of a few moments he repeated this action, and when I remarked that he probably resented the presence of a stranger, Melford exclaimed, "Oh, no, he wants to play with you--that's all."

His manner of playing was rather startling. So long as I kept my eyes on him he remained motionless, but the instant my attention wandered, or when in speaking I looked at my companion, the sudden violent dash at my head would be made.

I was assured by Melford that his birds never carried off and concealed bright objects, a habit which it has been said the jay, as well as the magpie, possesses.

"What would he do with this shilling if I tossed it to him?" I asked.

"Catch it," he returned. "It would simply be play to him, but he wouldn't carry it off."

I tossed up the shilling, and the bird had perhaps expected me to do so, as he deftly caught it just as a dog catches a biscuit when you toss one to him. After keeping it a few moments in his beak, he put it down at his side. I took out four

---

more shilling pieces and tossed them quickly one by one, and he caught them without a miss and placed them one by one with the other, not scattered about, but in a neat pile. Then, seeing that I had no more shillings he flew off.

After these few playful passages with one of his birds, I could understand Melford's feeling about his free pet jays, magpies and jackdaws; they were not merely birds to him, but rather like so many delightful little children in the beautiful shape of birds.

*

There was no rookery in or near the village, but a large flock of rooks were always to be seen feeding and sunning themselves in some level meadows near the river. It struck me one day as a very fine sight, when an old bird, who looked larger and blacker and greyer-faced than the others, and might have been the father and leader of them all, got up on a low post, and with wide-open beak poured forth a long series of most impressive caws. One always wonders at the meaning of such displays. Is the old bird ad-

BIRDS IN A VILLAGE 29

dressing the others in the rook language on some matter of great moment; or is he only expressing some feeling in the only language he has--those long, hoarse, uninflected sounds; and if so, what feeling? Probably a very common one. The rooks appeared happy and prosperous, feeding in the meadow grass in that June weather, with the hot sun shining on their glossy coats. Their days of want were long past and forgotten; the anxious breeding period was over; the tempest in the tall trees; the annual slaughter of the young birds--all past and forgotten. The old rook was simply expressing the old truth, that life was worth living.

These rooks were usually accompanied by two or three or more crows--a bird of so ill-repute that the most out-and-out enthusiast for protection must find it hard to say a word in its favour. At any rate, the rooks must think, if they think at all, that this frequent visitor and attendant of theirs is more kin than kind. I have related in a former work that I once saw a peregrine strike down and kill an owl--a sight that made me gasp with astonishment. But I am inclined to think of this act as only a slip, a slight aberration,

30 BIRDS IN TOWN AND VILLAGE

on the part of the falcon, so universal is the sense of relationship among the kinds that have the rapacious habit; or, at the worst, it was merely an isolated act of deviltry and daring of the sharp-winged pirate of the sky, a sudden assertion of over-mastering energy and power, and a very slight offence compared with that of the crow when he carries off and devours his callow little cousins of the rookery.

*

One of the first birds I went out to seek--perhaps the most medicinal of all birds to see--was the kingfisher; but he was not anywhere on the river margin, although suitable places were plentiful enough, and myriads of small fishes were visible in the shallow water, seen at rest like dim-pointed stripes beneath the surface, and darting away and scattering outwards, like a flight of arrows, at any person's approach. Walking along the river bank one day, when the place was still new to me, I discovered a stream, and following it up arrived at a spot where a clump of trees overhung the water, casting on it a deep shade. On the other side of the stream butter-

---

cups grew so thickly that the glazed petals of the flowers were touching; the meadow was one broad expanse of brilliant yellow. I had not been standing half a minute in the shade before the bird I had been seeking darted out from the margin, almost beneath my feet, and then, instead of flying up or down stream, sped like an arrow across the field of buttercups. It was a very bright day, and the bird going from me with the sunshine full on it, appeared entirely of a shining, splendid green. Never had I seen the kingfisher in such favourable circumstances; flying so low above the flowery level that the swiftly vibrating wings must have touched the yellow petals; he was like a waif from some far tropical land. The bird was tropical, but I doubt if there exists within the tropics anything to compare with a field of buttercups--such large and unbroken surfaces of the most brilliant colour in nature. The first bird's mate appeared a minute later, flying in the same direction, and producing the same splendid effect, and also green. These two alone were seen, and only on this occasion, although I often revisited the spot, hoping to find them again.

Now, the kingfisher is blue, and I am puzzled to know why, on this one occasion, it appeared green. I have, in a former work, *Argentine Ornithology*, described a contrary effect in a small and beautiful tyrantbird, *Cyanotis azarae*, variously called, in the vernacular, "All-colored or Many-colored Kinglet." It has a little blue on its head, but its entire back, from the nape to the tail, is deep green. It lives in beds of bulrushes, and when seen flying from the spectator in a very strong light, at a distance of twenty or thirty yards, its colour in appearance is bright cerulean blue. It is a sunlight effect, but how produced is a mystery to me. In the case of the two green kingfishers, I am inclined to think that the yellow of that shining field of buttercups in some way produced the illusion.

Why are these exquisite birds so rare, even in situations so favourable to them as the one I have described? Are they killed by severe frosts? An ornithological friend from Oxfordshire assures me that it will take several favourable seasons to make good the losses of the late terrible winter of 1891-92. But this, as every ornithologist knows, is only a part of the truth. The large

---

BIRDS IN A VILLAGE 33

number of stuffed kingfishers under glass shades that one sees in houses of all descriptions, in town and country, but most frequently in the parlours of country cottages and inns, tell a melancholy story. Some time ago a young man showed me three stuffed kingfishers in a case, and informed me that he had shot them at a place (which he named) quite close to London. He said that these three birds were the last of their kind ever seen there; that he had gone, week after week and watched and waited, until one by one, at long intervals, he had secured them all; and that two years had passed since the last one was killed, and no other kingfisher had been seen at the place. He added that the waterside which these birds had frequented was resorted to by crowds of London working people on Saturday afternoons, Sundays and other holidays; the fact that hundreds, perhaps thousands, of pairs of tired eyes would have been freshened and gladdened by the sight of their rare gem-like beauty only made him prouder of his achievement. This young man was a cockney of the small shop-keeping class--a Philistine of the Philistines--hence there was no call to feel surprise at his self-glori-

---

fication over such a matter. But what shall we say of that writer whose masterly works on English rural life are familiar to everyone, who is regarded as first among "lovers of nature," when he relates that he invariably carried a gun when out of doors, mainly with the object of shooting any kingfisher he might chance to see, as the dead bird always formed an acceptable present to the cottager's wife, who would get it stuffed and keep it as an ornament on her parlour mantelshelf!

Happily for the kingfisher, and for human beings who love nature, the old idea that beautiful birds were meant to be destroyed for fun by anyone and everyone, from the small-brained, detestable cockney sportsman I have mentioned, to the gentlemen who write books about the beauties of nature, is now gradually giving place to this new one--that it would be better to preserve the beautiful things we possess. Half a century before the author of "Wild Life in a Southern Country" amused himself by carrying a gun to shoot kingfishers, the inhabitants of that same county of Wiltshire were bathed in tears--so I read in an old Salisbury newspaper--at the tragic death of a young gentleman of great distinction,

great social charm, great promise. He was out shooting swallows with a friend who, firing at a passing swallow, had the misfortune to shoot and kill *him.*

At the present time when gentlemen practise a little at flying birds, to get their hand in before the first of September, they shoot sparrows as a rule, or if they shoot swallows, which afford them better practice, they do not say anything about it.

IV

WHERE the stream broadened and mixed with the river, there existed a dense and extensive rush-bed--an island of rushes separated by a deep channel, some twelve or fourteen yards in width from the bank. This was a favourite nesting-place of the sedge-warblers; occasionally as many as a dozen birds could be heard singing at the same time, although in no sense together, and the effect was indeed curious. This is not a song that spurts and gushes up fountain-like in the manner of the robin's, and of some other kinds, sprinkling the listener, so to speak, with a sparkling vocal spray; but it keeps low down, a song that flows along the surface gurgling and prattling like musical running water, in its shallow pebbly channel. Listening again, the similitude that seemed appropriate at first was cast aside for another, and then another still. The hidden

singers scattered all about their rushy island were small, fantastic, human minstrels, performing on a variety of instruments, some unknown, others recognizable--bones and castanets, tiny hurdy-gurdies, piccolos, banjos, tabours, and Pandean pipes--a strange medley!

Interesting as this concert was, it held me less than the solitary singing of a sedge-warbler that lived by himself, or with only his mate, higher up where the stream was narrow, so that I could get near him; for he not only tickled my ears with his rapid, reedy music, but amused my mind as well with a pretty little problem in bird psychology. I could sit within a few yards of his tangled haunt without hearing a note; but if I jumped up and made a noise, or struck the branches with my stick, he would incontinently burst into song. It is a very well-known habit of the bird, and on account of it and of the very peculiar character of the sounds emitted, his song is frequently described by ornithologists as "mocking, defiant, scolding, angry," *etc.* It seems clear that at different times the bird sings from different exciting causes. When, undisturbed by a strange presence, he bursts spontaneously into singing, the

N

38 BIRDS IN TOWN AND VILLAGE

music, as in other species, is simply an expression of overflowing gladness; at other times, the bird expressed such feelings as alarm, suspicion, solicitude, perhaps anger, by singing the same song. How does this come about?

I have stated, when speaking of the nightingale, that birds in which the singing faculty is highly developed, sometimes make the mistake of bursting into song when anxious or distressed or in pain, but that this is not the case with the mocking-birds. Some species of these brilliant songsters of the New World, in their passion for variety (to put it that way), import every harsh and grating cry and sound they know into their song; but, on the other hand, when anxious for the safety of their young, or otherwise distressed, they emit only the harsh and grating sounds--never a musical note. In the sedge-warbler, the harsh, scolding sounds that express alarm, solicitude, and other painful emotions, have also been made a part of the musical performance; but this differs from the songs of most species, the mocking birds included, in the extraordinary rapidity with which it is enunciated; once the song begins it goes on swiftly to the finish, harsh and

BIRDS IN A VILLAGE 39

melodious notes seeming to overlap and mingle, the sound forming, to speak in metaphor, a close intricate pattern of strongly-contrasted colours. Now the song invariably begins with the harsh notes--the sounds which, at other times, express alarm and other more or less painful emotions--and it strikes me as a probable explanation that when the bird in the singing season has been startled into uttering these harsh and grating sounds, as when a stone is flung into the rushes, he is incapable of uttering them only, but the singing notes they suggest and which he is in the habit of uttering, follow automatically.

The spot where I observed this wee feathered fantasy, the tantalizing sprite of the rushes, and where I soon ceased to see, hear, or think about him, calls for a fuller description. On one side the wooded hill sloped downward to the stream; on the other side spread the meadows where the rooks came every day to feed, or to sit and stand about motionless, looking like birds cut out of jet, scattered over about half an acre of the grassy, level ground. Stout old pollard willows grew here and there along the banks and were pleasant to see, this being the one man-mutilated thing in

---

nature which, to my mind, not infrequently gains in beauty by the mutilation, so admirably does it fit into and harmonize with the landscape. At one point there was a deep, nearly stagnant pool, separated from the stream by a strip of wet, rushy ground, its still dark surface covered with water-lilies, not yet in bloom. They were just beginning to show their polished buds, shaped like snake's heads, above the broad, oily leaves floating like islands on the surface. The stream itself was, on my side, fringed with bulrushes and other aquatic plants; on the opposite bank there were some large alders lifting their branches above great masses of bramble and rose-briar, all together forming as rich and beautiful a tangle as one could find even in the most luxuriant of the wild, unkept hedges round the village. The briars especially flourished wonderfully at this spot, climbing high and dropping their long, slim branches quite down to the surface of the water, and in some places forming an arch above the stream. A short distance from this tangle, so abundantly sprinkled with its pale delicate roses, the water was spanned by a small wooden bridge, which no person appeared to use, but which had

---

WREN.
"... mysterious talk in the leaves,"

BIRDS IN A VILLAGE 41

a use. It formed the one dry clear spot in the midst of all that moist vegetation, and the birds that came from the wood to drink and search for worms and small caterpillars first alighted on the bridge. There they would rest a few moments, take a look round, then fly to some favourite spot where succulent morsels had been picked up on previous visits.

Thrushes, blackbirds, sparrows, reed-buntings, chaffinches, tits, wrens, with many other species, succeeded each other all day long; for now they mostly had young to provide for, and it was their busiest time.

The unsullied beauty and solitariness of this spot made me wish at first that I was a boy once more, to climb and to swim, to revel in the sunshine and flowers, to be nearer in spirit to the birds and dragon flies and water-rats; then, that I could build a cabin and live there all the summer long, forgetful of the world and its affairs, with no human creature to keep me company, and no book to read, or with only one slim volume, some Spanish poet, let me say Melendez, for preference--only a small selection from his too voluminous writings; for he, albeit an eighteenth-century singer, was perhaps

---

the last of that long, illustrious line of poets who sang as no others have sung of the pure delightfulness of a life with nature. Something of this charm is undoubtedly due to the beauty of the language they wrote in and to the free, airy grace of assonants. What a hard, artificial sound the rhyme too often has: the clink that falls at regular intervals as of a stone-breaker's hammer! In the freer kinds of Spanish poetry there are numberless verses that make the smoothest lines and lyrics of our sweetest and most facile singers, from Herrick to Swinburne, seem hard and mechanical by comparison. But there is something more. I doubt, for one thing, if we are justified in the boast we sometimes make that the feeling for Nature is stronger in our poets than in those of other countries. The most scientific critic may be unable to pick a hole in Tennyson's botany and zoology; but the passion for, and feeling of oneness with Nature may exist without this modern minute accuracy. Be this as it may, it was not Tennyson, nor any other of our poets, that I would have taken to my dreamed of solitary cabin for companionship: Melendez came first to my mind. I think of his lines to a butterfly:

BIRDS IN A VILLAGE 43

De donde alegre vienes Tan suelta y tan festiva, Las valles alegrando Veloz mariposilla?*

and can imagine him--the poet himself--coming to see me through the woods and down the hill with the careless ease and lightness of heart of his own purple-winged child of earth and air--*tan suelta y tan festiva.* Here in these four or five words one may read the whole secret of his charm--the exquisite delicacy and seeming artlessness in the form, and the spirit that is in him--the old, simple, healthy, natural gladness in nature, and feeling of kinship with all the children of life. But I do not wish to disturb anyone in his prepossessions. It would greatly trouble me to think that my reader should, for the space of a page, or even of a single line, find himself in opposition to and not with me; and I am free

* May be roughly rendered thus:

Whence, blithe one, comest thou With that airy, happy flight--To make the valleys glad, O swift-winged butterfly?

---

44 BIRDS IN TOWN AND VILLAGE

to admit that with regard to poetry one's preferences change according to the mood one happens to be in and to the conditions generally. At home in murky London on most days I should probably seek pleasure and forgetfulness in Browning; but in such surroundings as I have been describing the lighter-hearted, elf-like Melendez accords best with my spirit, one whose finest songs are without human interest; who is irresponsible as the wind, and as unstained with earthly care as the limpid running water he delights in: who is brother to bird and bee and butterfly, and worships only liberty and sunshine, and is in love with nothing but a flower.

Nearly midway between the useful little bridge and the rose-blossoming tangle I have spoken of there were three elm-trees growing in the open grassy space near the brook; they were not lofty, but had very wide-spreading horizontal branches, which made them look like oaks. This was an ideal spot in which to spend the sultry hours, and I had no sooner cast myself on the short grass in the shade than I noticed that the end of a projecting branch above my head, and about twenty feet from the ground, was a favourite perch of

a tree-pipit. He sang in the air and, circling gracefully down, would alight on the branch, where, sitting near me and plainly visible, he would finish his song and renew it at intervals; then, leaving the loved perch, he would drop, singing, to the ground, just a few yards beyond the tree's shadow; thence, singing again, he would mount up and up above the tree, only to slide down once more with set, unfluttering wings, with a beautiful swaying motion to the same old resting-place on the branch, there to sing and sing and sing.

If Melendez himself had come to me with flushed face and laughing eyes, and sat down on the grass at my side to recite one of his most enchanting poems, I should, with finger on lip, have enjoined silence; for in the mood I was then in at that sequestered spot, with the landscape outside my shady green pavilion bathed and quivering in the brilliant sunshine, this small bird had suddenly become to me more than any other singer, feathered or human. And yet the tree-pipit is not very highly regarded among British melodists, on account of the little variety there is in its song. Nevertheless, it is most sweet--perhaps the

sweetest of all. It is true that there are thousands, nay, millions of things--sights and sounds and perfumes--which are or may be described as sweet, so common is the metaphor, and this too common use has perhaps somewhat degraded it; but in this case there is no other word so well suited to describe the sensation produced.

The tree-pipit has a comparatively short song, repeated, with some variation in the number and length of the notes, at brief intervals. The opening notes are thick and throaty, and similar in character to the throat-notes of many other species in this group, a softer sound than the throat-notes of the skylark and woodlark, which they somewhat resemble. The canary-like trills and thin piping notes, long drawn out, which follow vary greatly in different individuals, and in many cases the trills are omitted. But the concluding notes of the song I am considering--which is only one note repeated again and again--are clear and beautifully inflected, and have that quality of sweetness, of lusciousness, I have mentioned. The note is uttered with a downward fall, more slowly and expressively at each repetition, as if the singer felt overcome at the sweet-

---

ness of life and of his own expression, and languished somewhat at the close; its effect is like that of the perfume of the honeysuckle, infecting the mind with a soft, delicious languor, a wish to lie perfectly still and drink of the same sweetness again and again in larger measure.

To some who are familiar with this by no means uncommon little bird, it may seem that I am overstating the charm of its melody. I can only say that the mood I was then in made me very keenly appreciative; also that I have never heard any other individual of this species able to produce precisely the same effect. We know that there are quite remarkable differences in the songs of birds of the same species, that among several that appear to be perfect and to sing alike one will possess a charm above the other. The truth is they are not alike; they affect us differently, but the sense is not fine enough or not sufficiently trained to detect the cause. The poet's words may be used of this natural melody as well as of the works of art:

"O the little more and how much it is!"

TEere were about the village, within a few minutes' walk of the cottage, not fewer than half-a-dozen tree-pipits, each inhabiting a favourite spot where I could always count on finding and hearing him at almost any hour of the day from sunrise to sunset. Yet I cared not for these. To the one chosen bird I returned daily to spend the hot hours, lying in the shade and listening to his strain. Finally, I allowed two or three days to slip by, and when I revisited the old spot the secret charm had vanished. The bird was there, and rose and fell as formerly, pouring out his melody; but it was not the same: something was missing from those last sweet, languishing notes. Perhaps in the interval there had been some disturbing accident in his little wild life, though I could hardly believe it, since his mate was still sitting about thirty yards from the tree on the five little mottled eggs in her nest. Or perhaps his midsummer's music had reached its highest point, and was now in its declension. And perhaps the fault was in me. The virtue that draws and holds us does not hold us always, nor very long; it departs from all things, and we wonder why. The loss is in ourselves, although

we do not know it. Nature, the chosen mistress of our heart, does not change towards us, yet she is now, even to-day--

"Less full of purple colour and hid spice,"

and smiles and sparkles in vain to allure us, and when she touches us with her warm, caressing touch, there is, compared with yesterday, only a faint response.

V

COMING back from the waterside through the wood, after the hottest hours of the day were over, the crooning of the turtledoves would be heard again on every side--that summer beech-wood lullaby that seemed never to end. The other bird voices were of the willow-wren, the wood-wren, the coal-tit, and the now somewhat tiresome chiffchaff; from the distance would come the prolonged rich strain of the blackbird, and occasionally the lyric of the chaffinch. The song of this bird gains greatly when heard from a tall tree in the woodland silence; it has then a resonance and wildness which it appears to lack in the garden and orchard. In the village I had been glad to find that the chaffinch was not too common, that in the tangle of minstrelsy one could enjoy there his vigorous voice was not predominant.

Of all these woodland songsters the wood-wren impressed me the most. He could always be heard, no matter where I entered the wood, since all this world of tall beeches was a favoured haunt of the wood-wren, each pair keeping to its own territory of half-an-acre of trees or so, and somewhere among those trees the male was always singing, far up, invisible to eyes beneath, in the topmost sunlit foliage of the tall trees. On entering the wood I would, stand still for a few minutes to listen to the various sounds until that one fascinating sound would come to my ears from some distance away, and to that spot I would go to find a bed of last year's leaves to sit upon and listen. It was an enchanting experience to be there in that woodland twilight with the green cloud of leaves so far above me; to listen to the silence, to the faint whisper of the wind-touched leaves, then to little prelusive drops of musical sound, growing louder and falling faster until they ran into one prolonged trill. And there I would sit listening for half-an-hour or a whole hour; but the end would not come; the bird is indefatigable and with his mysterious talk in the leaves would tire the sun himself and send

52 BIRDS IN TOWN AND VILLAGE

him down the sky: for not until the sun has set and the wood has grown dark does the singing cease.

On emerging from the deep shade of the beeches into the wide grassy road that separated the wood from the orchards and plantations of fruit trees, and pausing for a minute to look down on the more than half-hidden village, invariably the first loud sounds that reached my ear were those of the cuckoo, thrush, and blackbird. At all hours in the village, from early morning to evening twilight, these three voices sounded far and near above the others. I considered myself fortunate that no large tree near the cottage had been made choice of by a song-thrush as a singing-stand during the early hours. The nearest tree so favoured was on the further side of a field, so that when I woke at half-past three or four o'clock, the shrill indefatigable voice came in at the open window, softened by distance and washed by the dewy atmosphere to greater purity. Throstle and skylark to be admired must be heard at a distance. But at that early hour when I sat by the open window, the cuckoo's call was the commonest sound; the birds were everywhere,

BIRDS IN A VILLAGE 53

bird answering bird far and near, so persistently repeating their double note that this sound, which is in character unlike any other sound in nature, which one so listens and longs to hear in spring, lost its old mystery and charm, and became of no more account than the cackle of the poultry-yard. It was the cuckoo's village; sometimes three or four birds in hot pursuit of each other would dash through the trees that lined the further side of the lane and alight on that small tree at the gate which the nightingale was accustomed to visit later in the day.

Other birds that kept themselves very much out of sight during most of the time also came to the same small tree at that early hour. It was regularly visited, and its thin bole industriously examined, by the nuthatch and the quaint little mouse-like creeper. Doubtless they imagined that five o'clock was too early for heavy human creatures to be awake, and were either ignorant of my presence or thought proper to ignore it.

But where, during the days when the vociferous cuckoo, with hoarse chuckle and dissyllabic call and wild bubbling cry was so much with us--where, in this period of many pleasant noises was

---

the cuckoo's mate, or maid, or messenger, the quaint and beautiful wryneck? There are few British birds, perhaps not one--not even the crafty black and white magpie, or mysterious moth-like goatsucker, or tropical kingfisher--more interesting to watch. At twilight I had lingered at the woodside, also in other likely places, and the goatsucker had failed to appear, gliding and zig-zagging hither and thither on his dusky-mottled noiseless wings, and now this still heavier disappointment was mine. I could not find the wryneck. Those quiet grassy orchards, shut in by straggling hedges, should have had him as a favoured summer guest. Creeper and nuthatch, and starling and gem-like blue tit, found holes enough in the old trunks to breed in. And yet I knew that, albeit not common, he was there; I could not exactly say where, but somewhere on the other side of the next hedge or field or orchard; for I heard his unmistakable cry, now on this hand, now on that. Day after day I followed the voice, sometimes in my eagerness forcing my way through a brambly hedge to emerge with scratched hands and clothes torn, like one that had been set upon and mauled by some

---

savage animal of the cat kind; and still the quaint figure eluded my vision.

At last I began to have doubts about the creature that emitted that strange, penetrating call. First heard as a bird-call, and nothing more, by degrees it grew more and more laugh-like--a long, far-reaching, ringing laugh; not the laugh I should like to hear from any person I take an interest in, but a laugh with all the gladness, unction, and humanity gone out of it--a dry mechanical sound, as if a soulless, lifeless, wind-instrument had laughed. It was very curious. Listening to it day by day, something of the strange history of the being once but no longer human, that uttered it grew up and took shape in my mind; for we all have in us something of this mysterious faculty. It was no bird, no wryneck, but a being that once, long, long, long ago, in that same beautiful place, had been a village boy--a free, careless, glad-hearted boy, like many another. But to this boy life was more than to others, since nature appeared immeasurably more vivid on account of his brighter senses; therefore his love of life and happiness in life greatly surpassed theirs. Annually the trees shed their

---

56 BIRDS IN TOWN AND VILLAGE

leaves, the flowers perished, the birds flew away to some distant country beyond the horizon, and the sun grew pale and cold in the sky; but the bright impression all things made on him gave him a joy that was perennial. The briony, woodbine, and honeysuckle he had looked on withered in the hedges, but their presentments flourished untouched by frost, as if his warmth sustained and gave them perpetual life; in that inner magical world of memory the birds still twittered and warbled, each after its kind, and the sun shone everlastingly. But he was living in a fool's paradise, as he discovered by-and-by, when a boy who had been his playmate began to grow thin and pale, and at last fell sick and died. He crept near and watched his dead companion lying motionless, unbreathing, with a face that was like white clay; and then, more horrible still, he saw him taken out and put into a grave, and the heavy, cold soil cast over him.

What did this strange and terrible thing mean? Now for the first time he was told that life is ours only for a season; that we also, like the leaves and flowers, flourish for a while then fade and perish, and mingle with the dust. The sad

---

knowledge had come too suddenly and in too vivid and dreadful a manner. He could not endure it. Only for a season!--only for a season! The earth would be green, and the sky blue, and the sun shine bright for ever, and he would not see, not know it! Struck with anguish at the thought, he stole away out of sight of the others to hide himself in woods and thickets, to brood alone on such a hateful destiny, and torture himself with vain longings, until he, too, grew pale and thin and large-eyed, like the boy that had died, and those who saw him shook their heads and whispered to one another that he was not long for this world. He knew what they were saying, and it only served to increase his misery and fear, and made him hate them because they were insensible to the awful fact that death awaited them, or so little concerned that they had never taken the trouble to inform him of it. To eat and drink and sleep was all they cared for, and they regarded death with indifference, because their dull sight did not recognize the beauty and glory of the earth, nor their dull hearts respond to Nature's everlasting gladness. The sight of the villagers, with their solemn head-

---

shakings and whisperings, even of his nearest kindred, grew insupportable, and he at length disappeared from among them, and was seen no more with his white, terror-stricken face. From that time he hid himself in the close thickets, supporting his miserable existence on wild fruits and leaves, and spending many hours each day lying in some sheltered spot, gazing up into that blue sunny sky, which was his to gaze on only for a season, while the large tears gathered in his eyes and rolled unheeded down his wasted cheeks.

At length during this period there occurred an event which is the obscurest part of his history; for I know not who or what it was--my mind being in a mist about it--that came to or accidentally found him lying on a bed of grass and dried leaves in his thorny hiding-place. It may have been a gipsy or a witch--there were witches in those days--who, suddenly looking on his upturned face and seeing the hunger in his unfathomable eyes, loved him, in spite of her malignant nature; or a spirit out of the earth; or only a very wise man, an ancient, white-haired solitary, whose life had been spent in finding out the secrets of nature. This being, becoming ac-

---

BIRDS IN A VILLAGE 59

quainted with the cause of the boy's grief and of his solitary, miserable condition, began to comfort him by telling him that no grief was incurable, no desire that heart could conceive unattainable. He discoursed of the hidden potent properties of nature, unknown only to those who seek not to know them; of the splendid virtue inherent in all things, like the green and violet flames in the clear colourless raindrops which are seen only on rare occasions. Of life and death, he said that life was of the spirit which never dies, that death meant only a passage, a change of abode of the spirit, and the left body crumbled to dust when the spirit went out of it to continue its existence elsewhere, but that those who hated the thought of such change could, by taking thought, prolong life and live for a thousand years, like the adder and tortoise or for ever. But no, he would not leave the poor boy to grope alone and blindly after that hidden knowledge he was burning to possess. He pitied him too much. The means were simple and near to hand, the earth teemed with the virtue that would save him from the dissolution which so appalled him. He would be startled to hear in

---

60 BIRDS IN TOWN AND VILLAGE

how small a thing and in how insignificant a creature resided the principle that could make his body, like his spirit, immortal. But exceeding great power often existed in small compass: witness the adder's tooth, which was to our sight no more than the point of the smallest thorn. Now, in the small ant there exists a principle of a greater potency than any other in nature; so strong and penetrating was it that even the dull and brutish kind of men who enquire not into hidden things know something of its power. But the greatest of all the many qualities of this acid was unknown to them. The ants were a small people, but exceedingly wise and powerful. If a little human child had the strength of an ant he would surpass in power the mightiest giant that ever lived. In the same way ants surpassed men in wisdom; and this strength and wisdom was the result of that acid principle in them. Now, if any person should be able to overcome his repugnance to so strange a food as to sustain himself on ants and nothing else, the effect of the acid on him would be to change and harden his flesh and make it impervious to decay or change of any

SONG THRUSH AND LONG-TAILED TIT.
" . . . a tree made choice of by a song-thrush as a singing-sland"

BIRDS IN A VILLAGE 61

kind. He would, so long as he confined himself to this kind of food, be immortal.

Not a moment did the wretched boy hesitatc to make use of this new and wonderful knowledge. When he had found and broken open an ant-hill, so eager was he that, shutting his eyes, he snatched up the maddened insects by handfuls and swallowed them, dust and ants together, and was then tortured for hours, feeling and thinking that they were still alive within him, running about in search of an outlet and frantically biting. The strange food sickened him, so that he grew thinner and paler, until at last he could barely crawl on hands and feet, and was like a skeleton except for the great sad eyes that could still see the green earth and blue sky, and still reflected in their depths one fear and one desire. And slowly, day by day, as his system accustomed itself to the new diet, his strength returned, and he was able once more to walk erect and run, and to climb a tree, where he could sit concealed among the thick foliage and survey the village where he had first seen the light and had passed the careless, happy years of boyhood. But he cherished no tender memories and regrets; his sole thought

---

was of the ants, and where to find a sufficiency of them to stay the cravings of hunger; for, after the first sensations of disgust had been overcome, he had begun to grow fond of this kind of food, and now consumed it with avidity. And as his strength increased so did his dexterity in catching the small, active insect prey. He no longer gathered the ants up in his palm and swallowed them along with dust and grit, but picked them up deftly, and conveyed them one by one to his mouth with lightning rapidity. Meanwhile that "acid principle," about which he had heard such wonderful things, was having its effect on his system. His skin changed its colour; he grew shrunken and small, until at length, after very many years, he dwindled to the grey little manikin of the present time. His mind, too, changed; he has no thought nor remembrance of his former life and condition and of his long-dead relations; but he still haunts the village where he knows so well where to find the small ants, to pick them from off the ant-hill and from the trunks of trees with his quick little claw-like hands. Language and song are likewise forgotten with all human things, all except his laugh; for

---

when hunger is satisfied, and the sun shines pleasantly as he reposes on the dry leaves on the ground or sits aloft on a branch, at times a sudden feeling of gladness possesses him, and he expresses it in that one way--the long, wild, ringing peal of laughter. Listening to that strange sound, although I could not see I could yet picture him, as, aware of my cautious approach, he moved shyly behind the mossy trunk of some tree and waited silently for me to pass. A lean, grey little man, clad in a quaintly barred and mottled mantle, woven by his own hands from some soft silky material, and a close-fitting brown peaked cap on his head with one barred feather in it for ornament, and a small wizened grey face with a thin sharp nose, puckered lips, and a pair of round, brilliant, startled eyes.

So distinct was this image to my mind's eye that it became unnecessary for me to see the creature, and I ceased to look for him; then all at once came disillusion, when one day, hearing the familiar high-pitched laugh with its penetrating and somewhat nasal tone, I looked and beheld the thing that had laughed just leaving its perch on a branch near the ground and winging

its way across the field. It was only a bird after all--only the wryneck; and that mysterious faculty I spoke of, saying that we all of us possessed something of it (meaning only some of us) was nothing after all but the old common faculty of imagination.

Later on I saw it again on half-a-dozen occasions, but never succeeded in getting what I call a satisfying sight of it, perched woodpecker-wise on a mossy trunk, busy at its old fascinating occupation of deftly picking off the running ants.

It is melancholy to think that this quaint and beautiful bird of a unique type has been growing less and less common in our country during the last half a century, or for a longer period. In the last fifteen or twenty years the falling-off has been very marked. The declension is not attributable to persecution in this case, since the bird is not on the gamekeeper's black list, nor has it yet become so rare as to cause the amateur collectors of dead birds throughout the country systematically to set about its extermination. Doubtless that will come later on when it will be in the same category with the golden oriole, hoopoe, furze-wren, and other species that are regarded as

---

always worth killing; that is to say, it will come--the scramble for the wryneck's carcass--if nothing is done in the meantime to restrain the enthusiasm of those who value a bird only when the spirit of life that gave it flight and grace and beauty has been crushed out of it--when it is no longer a bird. The cause of its decline up till now cannot be known to us; we can only say in our ignorance that this type, like innumerable others that have ceased to exist, has probably run its course and is dying out. Or it might be imagined that its system is undergoing some slow change, which tells on the migratory instinct, that it is becoming more a resident species in its winter home in Africa. But all conjectures are idle in such a case. It is melancholy, at all events for the ornithologist, to think of an England without a wryneck; but before that still distant day arrives let us hope that the love of birds will have become a common feeling in the mass of the population, and that the variety of our bird life will have been increased by the addition of some chance colonists and of many new species introduced from distant regions.

I have lingered long over the wryneck, but have still a story to relate of this bird--not a fairy tale this time, but true.

On the border of the village adjoining the wood--the side where birds were more abundant, and which consequently had the greatest attraction for me--there stands an old picturesque cottage nearly concealed from sight by the hedge in front and closely planted trees clustering round it. On one side was a grass field, on the other an orchard of old cherry, apple, and plum trees, all the property of the old man living in the cottage, who was a character in his way; at all events, he had not been fashioned in quite the same mould as the majority of the cottagers about him. They mostly, when past middle life, wore a heavy, dull and somewhat depressed look. This man had a twinkle in his dark-grey eyes, an expression of intelligent curiosity and fellowship; and his full face, bronzed with sixty or sixty-five years' exposure to the weather, was genial, as if the sunshine that had so long beaten on it had not been all used up in painting his skin that rich old-furniture colour, but had, some of it, filtered through the epidermis into the heart to make his

---

BIRDS IN A VILLAGE 67

existence pleasant and sweet. But it was a very rough-cast face, with shapeless nose and thick lips. He was short and broad-shouldered, always in the warm weather in his shirt-sleeves, a shirt of some very coarse material and of an earthen colour, his brown thick arms bare to the elbows. Waistcoat and trousers looked as if he had worn them for half his life, and had a marbled or mottled appearance as if they had taken the various tints of all the objects and materials he had handled or rubbed against in his life's work--wood, mossy trees, grass, clay, bricks, stone, rusty iron, and dozens more. He wore the field-labourer's thick boots; his ancient rusty felt hat had long lost its original shape; and finally, to complete the portrait, a short black clay pipe was never out of his lips--never, at all events, when I saw him, which was often; for every day as I strolled past his domain he would be on the outside of his hedge, or just coming out of his gate, invariably with something in his hand--a spade, a fork, or stick of wood, or an old empty fruit-basket. Although thus having the appearance of being very much occupied, he would always stop for a few minutes' talk with me; and by-and-by

---

68 BIRDS IN TOWN AND VILLAGE

I began to suspect that he was a very social sort of person, and that it pleased him to have a little chat, but that he liked to have me think that he met me by accident while going about his work.

One sunny morning as I came past his field he came out bearing a huge bundle of green grass on his head. "What!" he exclaimed, coming to a stand, "you here to-day? I thought you'd be away to the regatta."

I said that I knew little about regattas and cared less, that a day spent in watching and listening to the birds gave me more pleasure than all the regattas in the country. "I suppose you can't understand that?" I added.

He took the big green bundle from his head and set it down, pulled off his old hat to flap the dust out of it, then sucked at his short clay. "Well," he said at length, "some fancies one thing and some another, but we most of us like a regatta."

During the talk that followed I asked him if he knew the wryneck, and if it ever nested in his orchard. He did not know the bird; had never heard its name nor the other names of snake-bird and cuckoo's mate; and when I had min-

utely described its appearance, he said that no such bird was known in the village.

I assured him that he was mistaken, that I had heard the cry of the bird many times, and had even heard it once at a distance since our conversation began. Hearing that distant cry had caused me to ask the question.

All at once he remembered that he knew, or had known formerly, the wryneck very well, but he had never learnt its name. About twenty or five-and-twenty years ago, he said, he saw the bird I had just described in his orchard, and as it appeared day after day and had a strange appearance as it moved up the tree trunks, he began to be interested in it. One day he saw it fly into a hole close to the ground in an old apple tree. "Now I've got you!" he exclaimed, and running to the spot thrust his hand in as far as he could, but was unable to reach the bird. Then he conceived the idea of starving it out, and stopped up the hole with clay. The following day at the same hour he again put in his hand, and this time succeeded in taking the bird. So strange was it to him that after showing it to his own family he took it round to exhibit it to his neighbours,

---

and although some of them were old men, not one among them had ever seen its like before. They concluded that it was a kind of nuthatch, but unlike the common nuthatch which they knew. After they had all seen and handled it and had finished the discussions about it, he released it and saw it fly away; but, to his astonishment, it was back in his orchard a few hours later. In a few weeks it brought out its five or six young from the hole he had caught it in, and for several years it returned each season to breed in the same hole until the tree was blown down, after which the bird was seen no more.

What an experience the poor bird had suffered! First plastered up and left to starve or suffocate in its hollow tree; then captured and passed round from rough, horny hand to hand, while the villagers were discussing it in their slow, ponderous fashion--how wildly its little wild heart must have palpitated!--and, finally, after being released, to go back at once to its eggs in that dangerous tree. I do not know which surprised me most, the bird's action in returning to its nest after such inhospitable treatment, or the ignorance of the villagers concerning it. The incident

---

seemed to show that the wryneck had been scarce at this place for a very long period.

The villager, as a rule, is not a good observer, which is not strange, since no person is, or ever can be, a good observer of the things in which he is not specially interested; consequently the countryman only knows the most common and the most conspicuous species. He plods through life with downcast eyes and a vision somewhat dimmed by indifference;

forgetting, as he progresses, the small scraps of knowledge he acquired by looking sharply during the period of boyhood, when every living creature excited his attention. In Italy, notwithstanding the paucity of bird life, I believe that the peasants know their birds better. The reason of this is not far to seek; every bird, not excepting even the "temple-haunting martlet" and nightingale and minute golden-crested wren, is regarded only as a possible morsel to give a savour to a dish of polenta, if the shy, little flitting thing can only be enticed within touching distance of the limed twigs. Thus they take a very strong interest in, and, in a sense, "love" birds. It is their passion for this kind of flavouring which has drained rural Italy

---

72 BIRDS IN TOWN AND VILLAGE

of its songsters, and will in time have the same effect on Argentina, the country in which the withering stream of Italian emigration empties

itself.

VI

FROM the date of my arrival at the village in May, until I left it early in July, the great annual business of pairing, nest-building, and rearing the young was going on uninterruptedly. The young of some of the earliest breeders were already strong on the wing when I took my first walks along the hedgerows, still in their early, vivid green, frequently observing my bird through a white and rose-tinted cloud of apple-blossoms; and when I left some species that breed more than once in the season were rearing second broods or engaged in making new nests. On my very first day I discovered a nest full of fully fledged blue tits in a hole in an apple tree; this struck me as a dangerous place for the young birds; as the tree leaned over towards the lane, and the hole could almost be reached by a person standing on the ground. On the next day I went to look at them, and approaching noiselessly

along the lane, spied two small boys with bright clean faces--it was on a Sunday--standing within three or four yards of the tree, watching the tits with intense interest. The parent birds were darting up and down, careless of their presence, finding food so quickly in the gooseberry bushes growing near the roots of the tree that they visited the hole every few moments; while the young birds, ever screaming for more, were gathered in a dense little cluster at the entrance, their yellow breasts showing very brightly against the rain-wet wood and the dark interior of the hole. The instant the two little watchers caught sight of me the excited look vanished from their faces, and they began to move off, gazing straight ahead in a somewhat vacant manner. This instantaneous and instinctive display of hypocrisy was highly entertaining, and would have made me laugh if it had not been for the serious purpose I had in my mind. "Now, look here," I said, "I know what you are after, so it's no use pretending that you are walking about and seeing nothing in particular. You've been watching the young tits. Well, I've been watching them, too, and waiting to see them fly. I dare say they will

---

be out by to-morrow or the next day, and I hope you little fellows won't try to drag them out before then."

They at once protested that they had no such intention. They said that they never robbed birds' nests; that there were several nests at home in the garden and orchard, one of a nightingale with three eggs in it, but that they never took an egg. But some of the boys they knew, they said, took all the eggs they found; and there was one boy who got into every orchard and garden in the place, who was so sharp that few nests escaped him, and every nest he found he destroyed, breaking the eggs if there were any, and if there were young birds killing them.

Not, perhaps, without first mutilating them, I thought; for I know something of this kind of young "human devil," to use the phrase which Canon Wilberforce has made so famous in another connexion. Later on I heard much more about the exploits of this champion bird-destroyer of the village from (strange to say) a bird-catcher by trade, a man of a rather low type of countenance, and who lived, when at home, in a London slum. On the common where he spread his nets

---

he had found, he told me, about thirty nests containing eggs or fledglings; but this boy had gone over the ground after him, and not many of the nests had escaped his sharp eyes.

I was satisfied that the young tits were quite safe, so far as these youngsters were concerned, and only regretted that they were such small Boys, and that the great nest-destroyer, whose evil deeds they spoke of with an angry colour in their cheeks, was a very strong boy, otherwise I should have advised them to "go" for him.

Yes, he replied, some of the men would buy shot and use it early in the morning before their master was about; but if the man I had seen had been detected in the act, he would have been discharged on the spot. It was not only because the trees would be injured by shot, but this fruitgrower was friendly to birds.

Most fruitgrowers, I said, were dead against the birds, and anxious only to kill as many of them as possible.

It might be so in some places, he answered, but not in the village. He himself and most of the villagers depended, in a great measure, on the fruit they produced for a living, and their belief was that, taking one bird with another all the year round, the birds did them more good than harm.

I then imparted to him the views on this bird subject of a well-known fruitgrower in the north of England, Mr. Joseph Witherspoon, of Chester-le-Street. He began by persecuting the birds, as he had been taught to do by his father, a market-gardener; but after years of careful observation he completely changed his views, and is now so convinced of the advantage that birds are

to the fruitgrower, that he does all in his power to attract them, and to tempt them to breed in his grounds. His main idea is that birds that are fed on the premises, that live and feed among the trees, search for and attack the gardeners' enemies at every stage of their existence. At the same time he believes that it is very bad to grow fruit near woods, as in such a case the birds that live in the woods and are of no advantage to the garden, swarm into it as the fruit ripens, and that it is only by liberal use of nets that any reasonable portion of the fruit can be saved.

He answered that with regard to the last point he did not quite agree with Mr. Witherspoon. All the gardens and orchards in the village were raided by the birds from the wood, yet he reckoned they got as much fruit from their trees as others who had no woods near them. Then there was the big cherry plantation, one of the biggest in England, so that people came from all parts in the blossoming time just to look at it, and a wonderful sight it was. For a quarter of a mile this particular orchard ran parallel with the wood; with nothing but the green road between, and when the first fruit was ripening you could see

---

all the big trees on the edge of the wood swarming with birds--jays, thrushes, blackbirds, doves, and all sorts of tits and little birds, just waiting for a chance to pounce down and devour the cherries. The noise kept them off, but many would dodge in, and even if a gun was fired close to them the blackbirds would snatch a cherry and carry it off to the wood. That didn't matter--a few cherries here and there didn't count. The starlings were the worst robbers: if you didn't scare them they would strip a tree and even an orchard in a few hours. But they were the easiest birds to deal with: they went in flocks, and a shout or rattle or report of a gun sent the lot of them away together. His way of looking at it was this. In the fruit season, which lasts only a few weeks, you are bound to suffer from the attacks of birds, whether they are your own birds only or your own combined with others from outside, unless you keep them off; that those who do not keep them off are foolish or indolent, and deserve to suffer. The fruit season was, he said, always an anxious time.

In conclusion, I remarked that the means used for protecting the fruit, whether they served their

---

purpose well or not, struck me as being very unworthy of the times we lived in, and seemed to show that the British fruitgrowers, who were ahead of the world in all other matters connected with their vocation, had quite neglected this one point. A thousand years ago cultivators of the soil were scaring the birds from their crops just as we are doing, with methods no better and no worse, putting up scarecrows and old ragged garments and fluttering rags, hanging a dead crow to a stick to warn the others off, shouting and yelling and throwing stones. There appeared to be an opening here for experiment and invention. Mere noise was not terrifying to birds, and they soon discovered that an old hat on a stick had no injurious brains in or under it. But certain sounds and colours and odours had a strong effect on some animals. Sounds made to stimulate the screams of some hawks would perhaps prove very terrifying to thrushes and other small birds, and the effect of scarlet in large masses or long strips might be tried. It would also be worth while to try the effect of artificial sparrowhawks and other birds of prey, perched conspicuously, moving and perking their tails at intervals by

---

84 BIRDS IN TOWN AND VILLAGE

clockwork. In fact, a hundred things might be tried until something valuable was found, and when it lost its value, for the birds would in time discover the deception, some new plan adopted.

To this dissertation on what might be done, he answered that if any one could find out or invent any new effective means to keep the birds from the fruit, the fruitgrowers would be very thankful for it; but that no such invention could be looked for from those who are engaged on the soil; that it must come from those who do not dig and sweat, but sit still and work with their brains at new ideas.

This ended our conversation, and I left him more than satisfied at the information he had given me, and with a higher opinion than ever of his geniality and good practical sense.

It was a relief when the noisy, bird-scaring business was done with, and the last market baskets of ripe cherries were carried away to the station. Very splendid they looked in such large masses of crimson, as the baskets were brought out and set down in the grassy road; but I could not help thinking a little sadly that the thrushes and blackbirds which had been surreptitiously

BIRDS IN A VILLAGE 85

shot, when fallen and fluttering in the wet grass in the early morning, had shed life-drops of that same beautiful colour.

VII

AFTER the middle of June the common began to attract me more and more. It was so extensive that, standing on its border, just beyond the last straggling cottages and orchards, the further side was seen only as a line of blue trees, indistinct in the distance. As I grew to know it better, adding each day to my list from its varied bird life, the woods and waterside were visited less and less frequently, and after the bird-scaring noises began in the village, its wildness and quiet became increasingly grateful. The silence of nature was broken only by bird sounds, and the most frequent sound was that of the yellow bunting, as, perched motionless on the summit of a gorse bush, his yellow head conspicuous at a considerable distance, he emitted his thin monotonous chant at regular intervals, like a painted toy-bird that sings by machinery. There, too, sedentary as an owl in the daytime, the corn bunting was

common, discharging his brief song at intervals--a sound as of shattering glass. The whinchat was rarely seen, but I constantly met the small, prettily coloured stonechat flitting from bush to bush, following me, and never ceasing his low, querulous tacking chirp, anxious for the safety of his nest. Nightingales, blackcaps and white-throats also nested there, and were louder and more emphatic in their protests when approached. There were several grasshopper-warblers on the common, all, very curiously as it seemed to me, clustered at one spot, so that one could ramble over miles of ground without hearing their singular note; but on approaching the place they inhabited one gradually became conscious of a mysterious trilling buzz or whirr, low at first and growing louder and more stridulous, until the hidden singers were left behind, when by degrees it sank lower and lower again, and ceased to be audible at a distance of about one hundred yards from the points where it had sounded loudest. The birds hid in clumps of furze and bramble so near together that the area covered by the buzzing sound measured about two hundred yards across. This most singular sound (for a warbler

---

to make) is certainly not ventriloquial, although if one comes to it with the sense of hearing disorganized by town noises or unpractised, one is at a loss to determine the exact spot it comes from, or even to know from which side it comes. While emitting its prolonged sound the bird is so absorbed in its own performance that it is not easily alarmed, and will sometimes continue singing with a human listener standing within four or five yards of it. When one is near the bird, and listens, standing motionless, the effect on the nerves of hearing is very remarkable, considering the smallness of the sound, which, without being unpleasant, is somewhat similar to that produced by the vibration of the brake of a train; it is not powerful enough to jar the nerves, but appears to pervade the entire system. Lying still, with eyes closed, and three or four of these birds singing near, so that their strains overlap and leave no silent intervals, the listener can imagine that the sound originates within himself; that the numberless fine cords of his nervous network tremble responsively to it.

There are a number of natural sounds that resemble more or less closely the most unbird-

---

like note of this warbler--cicada, rattlesnake, and some batrachians. Some grasshoppers perhaps come nearest to it; but the most sustained current of sound emitted by the insect is short compared to the warbler's strain, also the vibrations are very much more rapid, and not heard as vibrations, and the same effect is not produced.

The grasshopper warblers gave me so much pleasure that I was often at the spot where they had their little colony of about half-a-dozen pairs, and where I discovered they bred every year. At first I used to go to any bush where I had caught sight of a bird and sit down within a few yards of it and wait until the little hideling's shyness wore off, and he would come out and start reeling. Afterwards I always went straight to the same bush, because I thought the bird that used it as his singing-place appeared less shy than the others. One day I spent a long time listening to this favourite; delightedly watching him, perched on a low twig on a level with my sight, and not more than five yards from me; his body perfectly motionless, but the head and wide-open beak jerked from side to side in a measured, mechanical way. I had a side view of the bird,

but every three seconds the head would be jerked towards me, showing the bright yellow colour of the open mouth. The reeling would last about three minutes, then the bird would unbend or unstiffen and take a few hops about the bush, then stiffen and begin again. While thus gazing and listening I, by chance, met with an experience of that rare kind which invariably strikes the observer of birds as strange and almost incredible--an example of the most perfect mimicry in a species which has its own distinctive song and is not a mimic except once in a while, and as it were by chance. The marsh warbler is our perfect mocking-bird, our one professional mimic; while the starling in comparison is but an amateur. We all know the starling's ever varying performance in which he attempts a hundred things and occasionally succeeds; but even the starling sometimes affects us with a mild astonishment, and I will here give one instance.

I was staying at a village in the Wiltshire downs, and at intervals, while sitting at work in my room on the ground floor, I heard the cackling of a fowl at the cottage opposite. I heard, but paid no attention to that familiar sound; but after three days it all at once struck me that no fowl could lay an egg about every ten or twelve minutes, and go on at this rate day after day, and, getting up, I went out to look for the cackler. A few hens were moving quietly about the open ground surrounding the cottage where the sound came from, but I heard nothing. By and by, when I was back in my room, the cackling sounded again, but when I got out the sound had ceased and the fowls, as before, appeared quite unexcited. The only way to solve the mystery was to stand there, out of doors, for ten minutes, and before that time was over a starling with a white grub in his beak, flew down and perched on the low garden wall of the cottage, then, with some difficulty, squeezed himself through a small opening into a cavity under a strip of zinc which covered the bricks of the wall. It was a queer place for a starling's nest, on a wall three feet high and within two yards of the cottage door which stood open all day. Having delivered the grub, the starling came out again and, hopping on to the zinc, opened his beak and cackled like a hen, then flew away for more grubs.

I observed the starling a good deal after this,

and found that invariably on leaving the nest, he uttered his imitation of a fowl cackling, and no other note or sound of any kind. It was as if he was not merely imitating a sound, but had seen a fowl leaving the nest and then cackling, and mimicked the whole proceeding, and had kept up the habit after the young were hatched.

To return to my experience on the common. About fifty yards from the spot where I was there was a dense thicket of furze and thorn, with a huge mound in the middle composed of a tangle of whitethorn and bramble bushes mixed with ivy and clematis. From this spot, at intervals of half a minute or so, there issued the call of a duck--the prolonged, hoarse call of a drake, two or three times repeated, evidently emitted in distress. I conjectured that it came from one of a small flock of ducks belonging to a cottage near the edge of the common on that side. The flock, as I had seen, was accustomed to go some distance from home, and I supposed that one of them, a drake, had got into that brambly thicket and could not make his way out. For half an hour I heard the calls without paying much attention, absorbed in watching the quaint little

---

songster close to me and his curious gestures when emitting his sustained reeling sounds. In the end the persistent distressed calling of the drake lost in a brambly labyrinth got a little on my nerves, and I felt it as a relief when it finally ceased. Then, after a short silence, another sound came from the same spot--a blackbird sound, known to everyone, but curiously interesting when uttered in the way I now heard it. It was the familiar loud chuckle, not emitted in alarm and soon ended, but the chuckle uttered occasionally by the bird when he is not disturbed, or when, after uttering it once for some real cause, he continues repeating it for no reason at all, producing the idea that he has just made the discovery that it is quite a musical sound and that he is repeating it, as if singing, just for pleasure. At such times the long series of notes do not come forth with a rush; he begins deliberately with a series of musical chirps uttered in a measured manner, like those of a wood wren, the prelude to its song, the notes coming faster and faster and swelling and running into the loud chuckling performance. This performance, like the lost drake's call, was repeated in the same deliberate

---

or leisurely manner at intervals again and again, until my curiosity was aroused and I went to the spot to get a look at the bird who had turned his alarm sound into a song and appeared to be very much taken with it. But there was no blackbird at the spot, and no lost drake, and no bird, except a throstle sitting motionless on the bush mound. This was the bird I had been listening to, uttering not his own thrush melody, which he perhaps did not know at all, but the sounds he had borrowed from two species so wide apart in their character and language.

The astonishing thing in this case was that the bird never uttered a note of his own original and exceedingly copious song; and I could only suppose that he had never learned the thrush melody; that he had, perhaps, been picked up as a fledgling and put in a cage, where he had imitated the sounds he heard and liked best, and made them his song, and that he had finally escaped or had been liberated.

The wild thrush, we know, does introduce certain imitations into his own song, but the borrowed notes, or even phrases, are, as a rule, few, and not always to be distinguished from his own.

BIRDS IN A VILLAGE 95

Sometimes one can pick them out; thus, on the borders of a marsh where redshanks bred, I have heard the call of that bird distinctly given by the thrush. And again, where the ring-ouzel is common, the thrush will get its brief song exactly. When thrushes taken from the nest are reared in towns, where they never hear the thrush or any other bird sing, they are often exceedingly vocal, and utter a medley of sounds which are sometimes distressing to the ear. I have heard many caged thrushes of this kind in London, but the most remarkable instance I have met with was at the little seaside town of Seaford. Here, in the main shopping street, a caged thrush lived for years in a butcher's shop, and poured out its song continuously, the most distressing throstle performance I ever heard, composed of a medley of loud, shrill and harsh sounds--imitations of screams and shouts, boy whistlers, saw filing, knives sharpened on steels, and numerous other unclassifiable noises; but all, more or less, painful. The whole street was filled with the noise, and the owner used to boast that his caged thrush was the most persistent as well as the loudest singer that had ever been heard. He had no

---

nerves, and was proud of it! On a recent visit to Seaford I failed to hear the bird when walking about the town, and after two or three days went into the shop to enquire about it. They told me it was dead--that it had been dead over a year; also that many visitors to Seaford had missed its song and had called at the shop to ask about the bird. The strangest thing about its end, they said, was its suddenness. The bird was singing its loudest one morning, and had been at it for some time, filling the whole place with its noise, when suddenly, in the middle of its song, it dropped down dead from its perch.

To drop dead while singing is not an unheard of, nor a very rare occurrence in caged birds, and it probably happens, too, in birds living their natural life. Listening to a nightingale, pouring out its powerful music continuously, as the lark sings, one sometimes wonders that something does not give way to end the vocalist's performance and life at the same instant. Some such incident was probably the origin of the old legend of the minstrel and the nightingale on which Strada based his famous poem, known in many languages. In England Crawshaw's version was by far

the best, and is perhaps the finest bird poem in our literature.

The blackbird, like the thrush, sometimes borrows a note or a phrase, and, like the thrush again, if reared by hand he may become a nuisance by mimicking some disagreeable sound, and using it by way of song. I heard of such a case a short time ago at Sidmouth. The ground floor of the house where I lodged was occupied by a gentleman who had a fondness for bird music, and being an invalid confined to his rooms, he kept a number of birds in cages. He had, besides canaries, the thrush, chaffinch, linnet, goldfinch and cirl bunting. I remarked that he did not have the best singer of all--the blackbird. He said that he had procured one, or that some friend had sent him one, a very beautiful ouel cock in the blackest plumage and with the orange-tawniest bill, and he had anticipated great pleasure from hearing its fluting melody. But alas! no blackbird song did this unnatural blackbird sing. He had learnt to bark like a dog, and whenever the singing spirit took him he would bark once or twice or three times, and then, after an interval of silence of the proper length, about

---

fifteen seconds, he would bark again, and so on until he had had his fill of music for the time. The barking got on the invalid's nerves, and he sent the bird away. "It was either that," he said, "or losing my senses altogether."

*

As all or most singing birds learn their songs from the adults of the same species, it is not strange that there should be a good deal of what we call mimicry in their performances: we may say, in fact, that pretty well all the true singers are mimics, but that some mimic more than others. Thus, the starling is more ready to borrow other birds' notes than the thrush, while the marsh-warbler borrows so much that his singing is mainly composed of borrowings. The nightingale is, perhaps, an exception. His voice excels in power and purity of sound, and what we may call his artistry is exceptionally perfect; this may account for the fact that he does not borrow from other birds' songs. I should say, from my own observation, that all songsters are interested in the singing of other species, or at all events,

---

in certain notes, especially the most striking in power, beauty, and strangeness. Thus, when the cuckoo starts calling, you will see other small birds fly straight to the tree and perch near him, apparently to listen. And among the listeners you will find the sparrow and tits of various species--birds which are never victimized by the cuckoo, and do not take him for a hawk since they take no notice of him until the calling begins. The reason that the double fluting call of the cuckoo is not mimicked by other birds is that they can't; because that peculiar sound is not in their register. The bubbling cry is reproduced by both the marsh warbler and the starling. Again, it is my experience that when a nightingale starts singing, the small birds near immediately become attentive, often suspending their own songs and some flying to perch near him, and listen, just as they listen to the cuckoo. Birds imitate the note or phrase that strikes them most, and is easiest to imitate, as when the thrush copies the piping and trilling of the redshank and the easy song of the ring-ouzel, which, when incorporated into his own music, harmonizes with it perfectly. But he cannot flute, and so never mimics the black-

bird's song, although he can and does, as we have seen, imitate its chuckling cry.

There is another thing to be considered. I believe that the bird, like creatures in other classes, has his receptive period, his time to learn, and that, like some mammals, he learns everything he needs to know in his first year or two; and that, having acquired his proper song, he adds little or nothing to it thereafter, although the song may increase in power and brilliance when the bird comes to full maturity. This, I think, holds true of all birds, like the nightingale, which have a singing period of two or three months and are songless for the rest of the year. That long, silent period cannot, so far as sounds go, be a receptive one; the song early in life has become crystallized in the form it will keep through life, and is like an intuitive act. This is not the case with birds like the starling, that sing all the year round--birds that are naturally loquacious and sing instead of screaming and chirping like others. They are always borrowing new sounds and always forgetting.

The most curious example of mimicry I have yet met with is that of a true mocking-bird,

Mimus patachonicus, a common resident species in northern Patagonia, on the Atlantic side, very abundant in places. He is a true mocking-bird because he belongs to the genus Mimus, a branch of the thrush family, and not because he mocks or mimics the songs of other species, like others of his kindred. He does not, in fact, mimic the set songs of others, although he often introduces notes and phrases borrowed from other species into his own performance. He sings in a sketchy way all the year round, but in spring has a fuller unbroken song, emitted with more power and passion. For the rest of the time he sings to amuse himself, as it seems, in a peculiarly leisurely, and one may say, indolent manner, perched on a bush, from time to time emitting a note or two, then a phrase which, if it pleases him, he will repeat two or three, or half a dozen times. Then, after a pause, other notes and phrases, and so on, pretty well all day long. This manner of singing is irritating, like the staccato song of our throstle, to a listener who wants a continuous stream of song; but it becomes exceedingly interesting when one discovers that the bird is thinking very much about his own music,

---

if one can use such an expression about a bird; that he is all the time experimenting, trying to get a new phrase, a new combination of the notes he knows and new notes. Also, that when sitting on his bush and uttering these careless chance sounds, he is, at the same time, intently listening to the others, all engaged in the same way, singing and listening. You will see them all about the place, each bird sitting motionless, like a grey and white image of a bird, on the summit of his own bush. For, although he is not gregarious as a rule, a number of pairs live near each other, and form a sort of loose community. The bond that unites them is their music, for not only do they sit within hearing distance, but they are perpetually mimicking each other. One may say that they are accomplished mimics but prefer mimicking their own to other species. But they only imitate the notes that take their fancy, so to speak. Thus, occasionally, one strikes out a phrase, a new expression, which appears to please him, and after a few moments he repeats it again, then again, and so on and on, and if you remain an hour within hearing he will perhaps be still repeating it at short intervals. Now, if

---

by chance there is something in the new phrase which pleases the listeners too, you will note that they instantly suspend their own singing, and for some little time they do nothing but listen. By and by the new note or phrase will be exactly reproduced from a bird on another bush; and he, too, will begin repeating it at short intervals. Then a second one will get it, then a third, and eventually all the birds in that thicket will have it. The constant repeating of the new note may then go on for hours, and it may last longer. You may return to the spot on the second day and sit for an hour or longer, listening, and still hear that same note constantly repeated until you are sick and tired of it, or it may even get on your nerves. I remember that on one occasion I avoided a certain thicket, one of my favourite daily haunts for three whole days, not to hear that one everlasting sound; then I returned and to my great relief the birds were all at their old game of composing, and not one uttered--perhaps he didn't dare--the too hackneyed phrase. I was sharply reminded one day by an incident in the village of this old Patagonian experience, and of the strange human-like weakness or pas-

---

sion for something new and arresting in music or song, something "tuney" or "catchy."

It chanced that when I left London a new popular song had come out and was "all the rage," a tune and words invented or first produced in the music-halls by a woman named Lottie Collins, with a chorus to it--*Ta-ra-ra-boom-de-ay*, repeated several times. First caught up in the music-halls it spread to the streets, and in ever-widening circles over all London, and over all the land. In London people were getting tired of hearing it, but when I arrived at my village "in a hole," and settled down among the Badgers, I heard it on every hand--in cottages, in the streets, in the fields, men, women and children were singing, whistling, and humming it, and in the evening at the inn roaring it out with as much zest as if they had been singing *Rule Britannia.*

This state of things lasted from May to the middle of June; then, one very hot, still day, about three o'clock, I was sitting at my cottage window when I caught the sound of a rumbling cart and a man singing. As the noise grew louder my interest in the approaching man and

---

cart was excited to an extraordinary degree; never had I heard such a noise! And no wonder, since the man was driving a heavy, springless farm cart in the most reckless manner, urging his two huge horses to a fast trot, then a gallop, up and down hill along those rough gully-like roads, he standing up in his cart and roaring out "Auld Lang Syne," at the top of a voice of tremendous power. He was probably tipsy, but it was not a bad voice, and the old familiar tune and words had an extraordinary effect in that still atmosphere. He passed my cottage, standing up, his legs wide apart, his cap on the back of his head, a big broad-chested young man, lashing his horses, and then for about two minutes or longer the thunder of the cart and the roaring song came back fainter, until it faded away in the distance. At that still hour of the day the children were all at school on the further side of the village; the men away in the fields; the women shut up in their cottages, perhaps sleeping. It seemed to me that I was the only person in the village who had witnessed and heard the passing of the big-voiced man and cart. But it was not so. At all events, next day, the whole village, men, women

---

and children, were singing, humming and whistling "Auld Lang Syne," and "Auld Lang Syne" lasted for several days, and from that day

"Ta-ra-ra-boom-de-ay" was heard no more. It had lost its charm.

VIII

JUST out of hearing of the grasshopper warblers, there was a good-sized pool of water on the common, probably an old gravel-pit, its bottom now overgrown with rushes. A sedge warbler, the only one on the common, lived in the masses of bramble and gorse on its banks; and birds of so many kinds came to it to drink and bathe that the pool became a favourite spot with me. One evening, just before sunset, as I lingered near it, a pied wagtail darted out of some low scrub at my feet and fluttered, as if wounded, over the turf for a space of ten or twelve yards before flying away. Not many minutes after seeing the wagtail, a reed-bunting--a bird which I had not previously observed on the common--flew down and alighted on a bush a few yards from me, holding a white crescent-shaped grub in its beak. I stood still to watch it, certainly not expecting to see its nest and young; for, as

a rule, a bird with food in its beak will sit quietly until the watcher loses patience and moves away; but on this occasion I had not been standing more than ten seconds before the bunting flew down to a small tuft of furze and was there greeted by the shrill, welcoming cries of its young. I went up softly to the spot, when out sprang the old bird I had seen, but only to drop to the ground just as the wagtail had done, to beat the turf with its wings, then to lie gasping for breath, then to flutter on a little further, until at last it rose up and flew to a bush.

After admiring the reed-bunting's action, I turned to the dwarf bush near my feet, and saw, perched on a twig in its centre, a solitary young bird, fully fledged but not yet capable of sustained flight. He did not recognise an enemy in me; on the contrary, when I approached my hand to him, he opened his yellow mouth wide, in expectation of being fed, although his throat was crammed with caterpillars, and the white crescent-shaped larva I had seen in the parent's bill was still lying in his mouth unswallowed. The wonder is that when a young bird had been stuffed with food to such an extent just before sleeping time,

he can still find it in him to open his mouth and call for more.

*

How wonderful it is that this parental instinct, so beautiful in its perfect simulation of the action of the bird that has lost the power of flight, should be found in so large a number of species! But when we find that it is not universal; that in two closely-allied species one will possess it and the other not; and that it is common in such widely-separated orders as gallinaceous and passerine birds, in pigeons, ducks, and waders, it becomes plain that it is not assignable to community of descent, but has originated independently all over the globe, in a vast number of species. Something of the beginnings and progressive development of this instinct may be learnt, I think, by noticing the behaviour of various passerine birds in the presence of danger, to their nests and young. Their actions and cries show that they are greatly agitated, and in a majority of species the parent bird flits and flutters round the intruder, uttering sounds of dis-

tress. Frequently the bird exhibits its agitation, not only by these cries and restless motions, but by the drooping of the wings and tail--the action observed in a bird when hurt or sick, or oppressed with heat. These languishing signs are common to a great many species after the young have been hatched; the period when the parental solicitude is most intense. In several species which I have observed in South America, the languishing is more marked. There are no sorrowful cries and restless movements; the bird sits with hanging wings and tail, gasping for breath with open bill--in appearance a greatly suffering bird. In some cases of this description, the bird, if it moves at all, hops or flutters from a higher to a lower branch, and, as if sick or wounded, seems about to sink to the ground. In still others, the bird actually does drop to the ground, then, feebly flapping its wings, rises again with great effort. From this last form it is but a step to the more highly developed complex instinct of the bird that sinks to the earth and flutters painfully away, gasping, and seemingly incapable of flight.

It would be a great mistake to suppose that the bird when fluttering on the ground to lead an

---

enemy from the neighbourhood of its nest is in full possession of all its faculties, acting consciously, and itself in as little danger of capture as when on its perch or flying through the air. We have seen that the action has its root in the bird's passion for its young, and intense solicitude in the presence of any danger threatening them, which is so universal in this class of creatures, and which expresses itself so variously in different kinds. This must be in all cases a painful and debilitating emotion, and when the bird drops down to the earth its pain has caused it to fall as surely as if it had received a wound or had been suddenly attacked by some grievous malady; and when it flutters on the ground it is for the moment incapable of flight, and its efforts to recover flight and safety cause it to beat its wings, and tremble, and gasp with open mouth. The object of the action is to deceive an enemy, or, to speak more correctly, the result is to deceive, and there is nothing that will more inflame and carry away any rapacious mammal than the sight of a fluttering bird. But in thus drawing upon itself the attention of an enemy threatening the safety of its eggs or young, to what a terrible

---

danger does the parent expose itself, and how often, in those moments of agitation and debility, must its own life fall a sacrifice! The sudden spring and rush of a feline enemy must have proved fatal in myriads of instances. From its inception to its most perfect stage, in the various species that possess it, this perilous instinct has been washed in blood and made bright.

What I have just said, that the peculiar instinct and deceptive action we have been considering is made and kept bright by being bathed in blood, applies to all instinctive acts that tend to the preservation of life, both of the individual and species. Necessarily so, seeing that, for one thing, instincts can only arise and grow to perfection in order to meet cases which commonly occur in the life of a species. The instinct is not prophetic and does not meet rare or extraordinary situations. Unless intelligence or some higher faculty comes in to supplement or to take the place of instinctive action then the creature must perish on account of the limitation of instinct. Again, the higher and more complete the instinct the more perilous it is, seeing that its efficiency depends on the absolutely perfect health and balance of all

---

the faculties and the entire organism. Thus, the higher instinctive faculty and action of birds for the preservation of the species, that of migration, is undoubtedly the most dangerous of all. It is so perfect that by means of this faculty millions and myriads of birds of an immense variety of species from cranes, swans, and geese down to minute goldcrests and firecrests and the smallest feeble-winged-leaf warblers, are able to inhabit and to distribute themselves evenly over all the temperate and cold regions of the earth, and even nearer the pole: and in all these regions they rear their young and spend several months each year, where they would inevitably perish from cold and lack of food if they stayed on to meet the winter. We can best realize the perfection of this instinct when we consider that all these migrants, including the young which have never hitherto strayed beyond the small area of their home where every tree and bush and spring and rock is familiar to them, rush suddenly away as if blown by a wind to unknown lands and continents beyond the seas to a distance of from a thousand to six or seven thousand miles; that after long months spent in those distant places, which in turn have grown

---

familiar to them, they return again to their natal place, not in a direct but ofttimes by a devious route, now north, now north-east, now east or west, keeping to the least perilous lines and crossing the seas where they are narrowest. Thus, when the returning multitude recrosses the Channel into England, coming by way of France and Spain from north or south or mid-Africa and from Asia, they at once proceed to disperse over the entire country from Land's End to Thurso and the northernmost islands of Scotland, until every wood and hill and moor and thicket and stream and every village and field and hedgerow and farmhouse has its own feathered people back in their old places. But they do not return in their old force. They had increased to twice or three times their original numbers when they left us, and as a result of that great adventure a half or two-thirds of the vast army has perished.

The instinct which in character comes nearest to that of the parent simulating the action of a wounded and terrified bird struggling to escape in order to safeguard its young, is that one, very strong in all ground-breeding species, of sitting close on the nest in the presence of danger. Here,

---

too, the instinct is of prime importance to the species, since the bird by quitting the nest reveals its existence to the prowling, nest-seeking enemy--dog, cat, fox, stoat, rat, in England; and in the country where I first observed animals, the skunk, armadillo, opossum, snake, wild cat, and animals of the weasel family. By leaving its nest a minute or half a minute too soon the bird sacrifices the eggs or young; by staying a moment too long it is in imminent danger of being destroyed itself. How often the bird stays too long on the nest is seen in the corn-crake, a species continually decreasing in this country owing to the destruction caused by the mowing-machine. The parent birds that escape may breed again in a safer place, but in many cases the bird clings too long to its nest and is decapitated or fatally injured by the cutters. Larks, too, often perish in the same way. To go back to the ailing or wounded bird simulating action: this is perhaps most perfect in the gallinaceous birds, all ground-breeders whose nests are most diligently hunted for by all egg-eating creatures, beast or bird, and whose tender chicks are a favourite food for all rapacious animals. In the fowl, pheasants, partridges, quail,

---

and grouse, the instinct is singularly powerful, the bird making such violent efforts to escape, with such an outcry, such beating of its wings and struggles on the ground, that no rapacious beast, however often he may have been deceived before, can fail to be carried away with the prospect of an immediate capture. The instinct and action has appeared to me more highly developed in these birds because, in the first place, the demonstrations are more violent than in other families, consequently more effective; and secondly, because the danger once over, the bird's recovery to its normal quiet, watchful state is quicker. By way of experiment, I have at various times thrown myself on pheasants, partridges and grouse, when I have found them with a family of recently-hatched chicks; then on giving up the chase and turning away from the bird its instantaneous recovery has seemed like a miracle. It was like a miracle because the creature did actually suffer from all those violent, debilitating emotions expressed in its disordered cries and action, and it is the miracle of Nature's marvellous health. If we, for example, were thrown into these violent extremes of passion, we should not escape the

---

after-effects. Our whole system would suffer, a doctor would perhaps have to be called in and would discourse wisely on metabolism and the development of toxins in the muscles, and give us a bottle of medicine.

I will conclude this digression and dissertation on a bird's instinct by relating the action of a hen-pheasant I once witnessed, partly because it is the most striking one I have met with of that instantaneous recovery of a bird from an extremity of distress and terror, and partly for another reason which will appear at the end.

The hen-pheasant was a solitary bird, having strayed away from the pheasant copses near the Itchen and found a nesting-place a mile away, on the other side of the valley, among the tall grasses and sedges on its border. I was the bird's only human neighbour, as I was staying in a fishing-cottage near the spot where the bird had its nest. Eventually, it brought off eight chicks and remained with them at the same spot on the edge of the valley, living like a rail among the sedges and tall valley herbage. I never went near the bird, but from the cottage caught sight of it from time to time, and sometimes watched it with my

binocular. There was, I thought, a good chance of its being able to rear its young, unless the damp proved injurious, as there was no dog or cat at the cottage, and there were no carrion crows or sparrowhawks at that spot. One morning about five o'clock on going out I spied a fox-terrier, a poaching dog from the neighbouring village, rushing about in an excited state a hundred yards or so below the cottage. He had scented the birds, and presently up rose the hen from the tall grass with a mighty noise, then flopping down she began beating her wings and struggling over the grass, uttering the most agonizing screams, the dog after her, frantically grabbing at her tail. I feared that he would catch her, and seizing a stick flew down to the rescue, yelling at the dog, but he was too excited to obey or even hear me. At length, thanks to the devious course taken by the bird, I got near enough to get in a good blow on the dog's back. He winced and went on as furiously as ever, and then I got in another blow so well delivered that the rascal yelled, and turning fled back to the village. Hot and panting from my exertions, I stood still, but sooner still the pheasant had pulled herself up and stood

BIRDS IN A VILLAGE 119

there, about three yards from my feet, as if nothing had happened--as if not a ripple had troubled the quiet surface of her life! The serenity of the bird, just out of that storm of violence and danger, and her perfect indifference to my presence, was astonishing to me. For a minute or two I stood still watching her; then turned to walk back to the cottage, and no sooner did I start than after me she came at a gentle trot, following me like a dog. On my way back I came to the very spot where the fox-terrier had found and attacked the bird, and at once on reaching it she came to a stop and uttered a call, and instantly from eight different places among the tall grasses the eight fluffy little chicks popped up and started running to her. And there she stood, gathering them about her with gentle chucklings, taking no notice of me, though I was standing still within two yards of her!

Up to the moment when the dog got his smart blow and fled from her she had been under the domination of a powerful instinct, and could have acted in no other way; but what guided her so infallibly in her subsequent actions? Certainly not instinct, and not reason, which hesitates between

120 BIRDS IN TOWN AND VILLAGE

different courses and is slow to arrive at a decision. One can only say that it was, or was like, intuition, which is as much as to say that we

don't know.

IX

AMONG the rarer fringilline birds on the common were the cirl bunting, bullfinch and goldfinch, the last two rarely seen. Linnets, however, were abundant, now gathered in small flocks composed mainly of young birds in plain plumage, with here and there an individual showing the carmine-tinted breast of the adult male. Unhappily, a dreary fate was in store for many of these blithe twitterers.

On June 24, when walking towards the pool, I spied two recumbent human figures on a stretch of level turf near its banks, and near them a something dark on the grass--a pair of clap-nets! "Still another serpent in my birds' paradise!" said I to myself, and, walking on, I skirted the nets and sat down on the grass beside the men. One was a rough brown-faced country lad; the other, who held the strings and wore the usual cap and comforter, was a man of about five-and-twenty,

---

with pale blue eyes and yellowish hair, close-cropped, and the unmistakable London mark in his chalky complexion. He regarded me with cold, suspicious looks, and, when I talked and questioned, answered briefly and somewhat surlily. I treated him to tobacco, and he smoked; but it wasn't shag, and didn't soften him. On mentioning casually that I had seen a stoat an hour before, he exhibited a sudden interest. It was as if one had said "rats!" to a terrier. I succeeded after a while in getting him to tell me the name of the man to whom he sent his captives, and when I told him that I knew the man well--a bird-seller in a low part of London--he thawed visibly. Finally I asked him to look at a red-backed shrike, perched on a bush about fifteen yards from his nets, through my field-glasses, and from that moment he became as friendly as possible, and conversed freely about his mystery. "How near it brings him!" he exclaimed, with a grin of delight, after looking at the bird. The shrike had greatly annoyed him; it had been hanging about for some time, he told me, dashing at the linnets and driving them off when they flew down to the nets. Two or three times he might have caught it, but would not draw the nets and have the trouble of resetting them for so worthless a bird. "But I'll take him the next time," he said vindictively. "I didn't know he was such a handsome bird." Unfortunately, the shrike soon flew away, and passing linnets dropped down, drawn to the spot by the twitterings of their caged fellows, and were caught; and so it went on for a couple of hours, we conversing amicably during the waiting intervals. For now he regarded me as a friend of the bird-catcher. Linnets only were caught, most of them young birds, which pleased him; for the young linnet after a month or two of cage life will sing; but the adult males would be silent until the next spring, consequently they were not worth so much, although the carmine stain in their breast made them for the time so much more beautiful.

I remarked incidentally that there were some who looked with unfriendly eyes on his occupation, and that, sooner or later, these people would try to get an Act of Parliament to make bird-catching in lanes, on commons and waste lands illegal. "They can't do it!" he exclaimed excitedly. "And if they can do it, and if they do

do it, it will be the ruination of England. For what would there be, then, to stop the birds increasing? It stands to reason that the whole country would be eaten up."

Doubtless the man really believed that but for the laborious days that bird-catchers spend lying on the grass, the human race would be very badly off.

Just after he had finished his protest, three or four linnets flew down and were caught. Taking them from the nets, he showed them to me, remarking, with a short laugh, that they were all young males. Then he thrust them down the stocking-leg which served as an entrance to the covered box he kept his birds in--the black hole in which their captive life begins, where they were now all vainly fluttering to get out. Going back to the previous subject, he said that he knew very well that many persons disliked a bird-catcher, but there was one thing that nobody could say against him--he wasn't cruel; he caught, but didn't kill. He only killed when he caught a great number of female linnets, which were not worth sending up; he pulled their heads off, and took them home to make a linnet pie. Then, by

---

way of contrast to his own merciful temper, he told me of the young nest-destroyer I have written about. It made him mad to see such things! Something ought to be done, he said, to stop a boy like that; for by destroying so many nestlings he was taking the bread out of the bird-catcher's mouth. Passing to other subjects, he said that so far he had caught nothing but linnets on the common--you couldn't expect to catch other kinds in June. Later on, in August and September, there would be a variety. But he had small hopes of catching goldfinches, they were too scarce now. Greenfinches, yellow-hammers, common buntings, reed sparrows--all such birds were worth only tuppence apiece. Oh, yes, he caught them just the same, and sent them up to London, but that was all they were worth to him. For young male linnets he got eightpence, sometimes tenpence; for hen birds fourpence, or less. I dare say that eightpence was what he hoped to get, seeing that young male linnets are not unfrequently sold by London dealers for sixpence and even fourpence. Goldfinches ran to eighteenpence, sometimes as much as two shillings. Starlings he had made a lot out of, but that was all past and over. Why?

Because they were not wanted--because people were such fools that they now preferred to shoot at pigeons. He hated pigeons! Gentlemen used to shoot starlings at matches; and if you had the making of a bird to shoot at, you couldn't get a better than the starling--such a neat bird! He had caught hundreds--thousands--and had sold them well. But now nothing but pigeons would they have. Pigeons! Always pigeons! He caught starlings still, but what was the good of that? The dealers would only take a few, and they were worth nothing--no more than greenfinches and yellow-hammers.

My colloquy with my enemy on the common tempts me to a fresh digression in this place--to have my say on a question about which much has already been said during the last three or four decades, especially during the 'sixties, when the first practical efforts to save our wild-bird life from destruction were made.

There is a feeling in the great mass of people that the pursuit of any wild animal, whether fit for food or not, for pleasure or gain, is a form of sport, and that sport ought not to be inter-

BIRDS IN A VILLAGE 127

fered with. So strong and well-nigh universal is this feeling, which is like a superstition, that the pursuit is not interfered with, however unsportsmanlike it may be, and when illegal, and when practised by only a very few persons in any district, where to others it may be secretly distasteful or even prejudicial.

Even bird-catching on a common is regarded as a form of sport and the bird-catcher as a sportsman--and a brother.

A striking instance of this tameness and stupidly acquiescent spirit in people generally was witnessed during the intensely severe frosts of the early part of the late winter (1882-3), when incalculable numbers of sea-birds were driven by hunger and cold into bays and inland waters. At this time thousands of gulls made their appearance in the Thames, but no sooner did they arrive than those who possessed guns and licences to shoot began to shoot them. The police interfered and some of these sportsmen were brought before the magistrates and fined for the offence of discharging guns to the public danger. For upwards of a fortnight after the shooting had been put a stop to, the gulls continued to frequent the river

---

in large numbers, and were perhaps most numerous from London Bridge to Battersea, and during this time they were watched every day by thousands of Londoners with keen interest and pleasure. The river here, flowing through the very centre and heart of the greatest city of the world, forms at all hours and at all seasons of the year a noble and magnificent sight; to my eyes it never looked more beautiful and wonderful than during those intensely cold days of January, when there was nothing that one could call a mist in a chilly, motionless atmosphere, but only a faint haze, a pallor as of impalpable frost, which made the heavens seem more white than blue, and gave a hoariness and cloud-like remoteness to the arches spanning the water, and the vast buildings on either side, ending with the sublime dome of the city cathedral; and when out of the pale motionless haze, singly, in twos and threes, in dozens and scores, floated the mysterious white bird-figures, first seen like vague shadows in the sky, then quickly taking shape and whiteness, and floating serenely past, to be succeeded by others and yet others.

It was not merely the ornithologist in me that

---

made the sight so fascinating, since it was found that others--all others, it might almost be said,--experienced the same kind of delight. Crowds of people came down to the river to watch the birds; workmen when released from their work at mid-day hurried down to the embankment so as to enjoy seeing the gulls while eating their dinners, and, strangest thing of all, to feed them with the fragments!

And yet these very men who found so great a pleasure in observing and feeding their white visitors from the sea, and were exhilarated with the novel experiences of seeing wild nature face to face at their own doors--these thousands would have stood by silent and consenting if the half-a-dozen scoundrels with guns and fish-hooks on lines had been allowed to have their will and had slaughtered and driven the birds from the river! And this, in fact, is precisely what happened at a distance from London, where guns could be discharged without danger to the public, in numberless bays and rivers in which the birds sought refuge. They were simply slaughtered wholesale in the most wanton manner; in More-cambe Bay a hundred and twelve gulls were killed

---

at one discharge, and no hand and no voice was raised to interfere with the hideous sport. Not because it was not shocking to the spectators, but because it was "Sport."

Doubtless it will be said that this wholesale wanton destruction of bird life, however painful it may be to lovers of nature, however reprehensible from a moral point of view, is sanctioned by law, and cannot therefore be prevented. This is not quite so. We see that the Wild Birds Protection Act is continually being broken with impunity, and where public opinion is unfavourable to it the guardians of the law themselves, the police and the magistrates, are found encouraging the people to break the law. Again, we find that where commons are enclosed, and the law says nothing, the people are accustomed to assemble together unlawfully to tear the fences down, and are not punished. For, after all, if laws do not express or square with public will or opinion, they have little force; and if, in any locality, the people thought proper to do so--if they were not restrained by that dull, tame spirit I have spoken of--they would, lawfully or unlawfully, protect their sea-fowl from the cockney

---

sportsmen, and sweep the bird-catchers out of their lanes and waste lands.

One day I paid a visit to Maidenhead, a pleasant town on the Thames, where the Thames is most beautiful, set in the midst of a rich and diversified country which should be a bird's paradise. In my walks in the town, I saw a great many stuffed kingfishers, and, in the shops of the local taxidermists, some rare and beautiful birds, with others that are fast becoming rare. But outside of the town I saw no kingfishers and no rare species at all, and comparatively few birds of any kind. It might have been a town of Philistine cockneys who at no very distant period had emigrated thither from the parish of St. Giles-in-the-Fields. I came home with the local guidebook in my pocket. It is now before me, and this is what its writer says of the Thicket, the extensive and beautiful common two miles from the town, which belongs to Maidenhead, or, in other words, to its inhabitants: "The Thicket was formerly much infested by robbers and highwaymen. The only remains of them to be found now are the snarers of the little feathered songsters, who im-

prison them in tiny cages and carry them off in large numbers to brighten by their sweet, sad sighs for liberty the dwellers in our smoky cities."

On this point I consulted a bird-catcher, who had spread his nets on the common for many years, and he complained bitterly of the increasing scarcity of its bird life. There was no better place than the Thicket formerly, he said; but now he could hardly make his bread there. I presume that a dozen men of his trade would be well able to drain the country in the neighbourhood of the Thicket of the greater portion of its bird life each year so as to keep the songsters scarce. Will any person maintain for a moment that the eight or nine thousand inhabitants of Maidenhead, and the hundreds or thousands inhabiting the surrounding country could not protect their songbirds from these few men, most of them out of London slums, if they wished or had the spirit to do so?

It is true that the local authorities in some country towns have made by-laws to protect the birds in their open spaces. Thus, at Tunbridge

Wells, since 1890, bird-trapping and bird's-nesting have been prohibited on the large and beau-

---

tiful common there; but, so far as I know, such measures have only been taken in boroughs after the birds have been almost exterminated.

Doubtless the day will come when, law or no law, the bird-catcher will find it necessary to go warily, lest the people of any place where he may be tempted to spread his nets should have formed the custom of treating those of his calling somewhat roughly. That it will come soon is earnestly to be wished. Nevertheless, it would be irrational to cherish feelings of animosity and hatred against the bird-catcher himself, the "man and brother," ready and anxious as we may be to take the bread out of his mouth. He certainly does not regard himself as an injurious or disreputable person; on the contrary he looks on himself as a useful member of the community, and in some cases even more. If anyone is to be hated or blamed, it is the person who sends the bird-catcher into the fields; not the dealer, but he who buys trapped birds and keeps them in cages to be amused by their twitterings. This is not a question of morality, nor of sentimentality, as some may imagine; but rather of taste, of the sense of fitness, of that something vaguely

---

described as the feeling for nature, which is not universal. Thus, one man will dine with zest on a pheasant, partridge, or quail, but would be choked by a lark; while another man will eat pheasant and lark with equal pleasure. Both may be good, honest, moral men; only one has that something which the other lacks. In one the soul responds to the skylark's music "singing at heaven's gate," in the other not; to one the roasted lark is merely a savoury morsel; the other, be he never so hungry, cannot dissociate the bird on the dish from that heavenly melody which registered a sensation in his brain, to be thereafter reproduced at will, together with the revived emotion. It is a curious question, and is no nearer to a settlement when one of these two I have described turns round and calls his neighbour a gross feeder, a worshipper of his belly, a soulless and brutish man; and when the other answers "pooh-pooh" and goes on complacently devouring larks with great gusto, until he is himself devoured of death.

To those with whom I am in sympathy in this matter, who love to listen to and are yearly invigorated by the skylark's music, and whose souls

---

are yearly sickened at the slaughter of their loved songsters, I would humbly suggest that there is a simpler, more practical means of ending this dispute, which has surely lasted long enough. It goes without saying that this bird's music is eminently pleasing to most persons, that even as the sunshine is sweet and pleasant to behold, its silvery aerial sounds rained down so abundantly from heaven are delightful and exhilarating to all of us, or at all events, to so large a majority that the minority are not entitled to consideration. One person in five thousand, or perhaps in ten thousand, might be found to say that the lark singing in blue heaven affords him no pleasure. This being so, and ours being a democratic country in which the will or desire of the many is or may be made the law of the land, it is surely only right and reasonable that lovers of lark's flesh should be prevented from gratifying their taste at the cost of the destruction of so loved a bird, that they should be made to content themselves with woodcock, and snipe on toast, and golden plover, and grouse and blackcock, and any other bird of delicate flavor which does not,

---

living, appeal so strongly to the aesthetic feelings in us and is not so universal a favourite.

This, too, will doubtless come in time. Speaking for myself, and going back to the former subject, little as I like to see men feeding on larks, rather would I see larks killed and eaten than thrust into cages. For in captivity they do not "sweeten" my life, as the Maidenhead guidebook writer would say, with their shrill, piercing cries for liberty, but they "sing me mad." Just as in some minds this bird's music--a sound which above all others typifies the exuberant life and joy of nature to the soul--cannot be separated from the cooked and dished-up melodist, so that they turn with horror from such meat, so I cannot separate this bird, nor any bird, from the bird's wild life of liberty, and the marvellous faculty of flight which is the bird's attribute. To see so wild and aerial a creature in a cage jars my whole system, and is a sight hateful and unnatural, an outrage on our universal mother.

This feeling about birds in captivity, which I have attempted to describe, and which, I repeat, is not sentimentality, as that word is ordinarily understood, has been so vividly rendered in an

---

BIRDS IN A VILLAGE 137

ode to "The Skylarks" by Sir Rennell Rodd, that the reader will probably feel grateful to me for quoting a portion of it in this place, especially as the volume in which it appears--*Feda, with Other Poems*--is, I imagine, not very widely known:

"Oh, the sky, the sky, the open sky,
For the home of a songbird's heart!
And why, and why, and for ever why,
Do they stifle here in the mart:
Cages of agony, rows on rows,
Torture that only a wild thing knows:
Is it nothing to you to see
That head thrust out through the hopeless wire,
And the tiny life, and the mad desire
To be free, to be free, to be free?

Oh, the sky, the sky, the blue, wide sky,
For the beat of a songbird's wings!

*

Straight and close are the cramping bars
From the dawn of mist to the chill of stars,
And yet it must sing or die!

Will its marred harsh voice in the city street
Make any heart of you glad?
It will only beat with its wings and beat,
It will only sing you mad.

*

If it does not go to your heart to see
The helpless pity of those bruised wings,
The tireless effort to which it clings
To the strain and the will to be free,
I know not how I shall set in words
The meaning of God in this,
For the loveliest thing in this world of His
Are the ways and the songs of birds.
But the sky, the sky, the wide, free sky,
For the home of the songbird's heart!"

* * * * *

How falsely does that man see Nature, how grossly ignorant must he be of its most elemental truths, who looks upon it as a chamber of torture, a physiological laboratory on a very vast scale, a scene of endless strife and trepidation, of hunger and cold, and every form of pain and misery--and who, holding this doctrine of

SKYLARK.
"; Oh, the sky, the sky, the open sky is the home of a songbird's heart,"

BIRDS IN A VILLAGE 139

Nature's cruelty, keeps a few captive birds in cages, and is accustomed to say of them, "These, at any rate, are safe, rescued from subjection to ruthless conditions, sheltered from the inclement weather and from enemies, and all their small wants abundantly satisfied;" who once or twice every day looks at his little captives, presents them with a lump of sugar, whistles and chuckles to provoke them to sing, then goes about his business, flattering himself that he is a lover of birds, a being of a sweet and kindly nature. It is all a delusion--a distortion and inversion of the truth--so absurd that it would be laughable were it not so sad, and the cause of so much unconscious cruelty. The truth is, that if birds be capable of misery, it is only in the unnatural conditions of a caged life that they experience it; and that if they are capable of happiness in a cage, such happiness or contentment is but a poor, pale emotion compared with the wild exuberant gladness they have in freedom, where all their instincts have full play, and where the perils that surround them do but brighten their many splendid faculties. The little bird twitters and sings in its cage, and among ourselves the blind man and the cripple

---

whistle and sing, too, feeling at times a lower kind of contentment and cheerfulness. The chaffinch in East London, with its eyeballs seared by red-hot needles, sings, too, in its prison, when it has grown accustomed to its darkened existence, and is in health, and the agreeable sensations that accompany health prompt it at intervals to melody, but no person, not even the dullest ruffian among the baser sort of bird-fanciers would maintain for a moment that the happiness of the little sightless captive, whether vocal or silent, is at all comparable in degree to that of the chaffinch singing in April "on the orchard bough," vividly seeing the wide sunlit world, blue above and green below, possessing the will and the power, when its lyric ends, to transport itself swiftly through the crystal fields of air to other trees and other woods.

I take it that in the lower animals misery can result from two causes only--restraint and disease; consequently, that animals in a state of nature are not miserable. They are not hindered nor held back. Whether the animal is migrating, or burying himself in his hibernating nest or den; or flying from some rapacious enemy, which he

---

may, or may not, be able to escape; or feeding, or sleeping, or fighting, or courting, or incubating, however many days or weeks this process may last--in all things he is obeying the impulse that is strongest in him at the time--he is doing what he wants to do--the one thing that makes him happy.

As to disease, it is so rare in wild animals, or in a large majority of cases so quickly proves fatal, that, compared with what we call disease in our own species it is practically non-existent. The "struggle for existence," in so far as animals in a state of nature are concerned, is a metaphorical struggle; and the strife, short and sharp, which is so common in nature, is not misery, although it results in pain, since it is pain that kills or is soon outlived. Fear there is, just as in fine weather there are clouds in the sky; and just as the shadow of the cloud passes, so does fear pass from the wild creature when the object that excited it has vanished from sight. And when death comes, it comes unexpectedly, and is not the death that we know, even before we taste of it, thinking of it with apprehension all our lives long, but a sudden blow that takes away consciousness--the

---

touch of something that numbs the nerves--merely the prick of a needle. In whatever way the animal perishes, whether by violence, or excessive cold, or decay, his death is a comparatively easy one. So long as he is fighting with or struggling to escape from an enemy, wounds are not felt as wounds, and scarcely hurt him--as we know from our own experience; and when overcome, if death be not practically instantaneous, as in the case of a small bird seized by a cat, the disabling grip or blow is itself a kind of anodyne, producing insensibility to pain. This, too, is a matter of human experience. To say nothing of those who fall in battle, men have often been struck down and fearfully lacerated by lions, tigers, jaguars, and other savage beasts; and after having been rescued by their companions, have recounted this strange thing. Even when there was no loss of consciousness, when they saw and knew that the animal was rending their flesh, they seemed not to feel it, and were, at the time, indifferent to the fate that had overtaken them.

It is the same in death from cold. The strong, well-nourished man, overtaken by a snowstorm on some pathless, uninhabited waste, may expe-

---

rience some exceedingly bitter moments, or even hours, before he gives up the struggle. The physical pain is simply nothing: the whole bitterness is in the thought that he must die. The horror at the thought of annihilation, the remembrance of all the happiness he is now about to lose, of dear friends, of those whose lives will be dimmed with grief for his loss, of all his cherished dreams of the future--the sting of all this is so sharp that, compared with it, the creeping coldness in his blood is nothing more than a slight discomfort, and is scarcely felt. By and by he is overcome by drowsiness, and ceases to struggle; the torturing visions fade from his mind, and his only thought is to lie down and sleep. And when he sleeps he passes away; very easily, very painlessly, for the pain was of the mind, and was over long before death ensued.

The bird, however hard the frost may be, flies briskly to its customary roosting-place, and with beak tucked into its wing, falls asleep. It has no apprehensions; only the hot blood grows colder and colder, the pulse feebler as it sleeps, and at midnight, or in the early morning, it drops from its perch--dead.

Yesterday he lived and moved, responsible to a thousand external influences, reflecting earth and sky in his small brilliant brain as in a looking-glass; also he had a various language, the inherited knowledge of his race, and the faculty of flight, by means of which he could shoot, meteor-like, across the sky, and pass swiftly from place to place, and with it such perfect control over all his organs, such marvellous certitude in all his motions, as to be able to drop himself plumb down from the tallest tree-top or out of the void air, on to a slender spray, and scarcely cause its leaves to tremble. Now, on this morning, he lies stiff and motionless; if you were to take him up and drop him from your hand, he would fall to the ground like a stone or a lump of clay--so easy and swift is the passage from life to death in wild nature! But he was never miserable.

Those of my readers who have seen much of animals in a state of nature, will agree that death from decay, or old age, is very rare among them. In that state the fullest vigour, with brightness of all the faculties, is so important that probably in ninety-nine cases in a hundred any falling-off in

---

strength, or decay of any sense, results in some fatal accident. Death by misadventure, as we call it, is Nature's ordinance, the end designed for a very large majority of her children. Nevertheless, animals do sometimes live on without accident to the very end of their term, to fade peacefully away at the last. I have myself witnessed such cases in mammals and birds; and one such case, which profoundly impressed me, and is vividly remembered, I will describe.

One morning in the late summer, while walking in the fields at my home in South America, I noticed a few purple martins, large, beautiful swallows common in that region, engaged, at a considerable height, in the aerial exercises in which they pass so much of their time each day. By and by, one of the birds separated itself from the others, and, circling slowly downward, finally alighted on the ground not far from me. I walked on: but the action of the bird had struck me as unusual and strange, and before going far, I turned and walked back to the spot where it continued sitting on the ground, quite motionless. It made no movement when I approached to within four yards of it; and after I had stood still

---

146 BIRDS IN TOWN AND VILLAGE

at that distance for a minute or so, attentively regarding it, I saw it put out one wing and turn over on its side. I at once took it up in my hand, and found that it was already quite dead. It was a large example of its species, and its size, together with a something of dimness in the glossy purple colour of the upper plumage, seemed to show that it was an old bird. But it was uninjured, and when I dissected it no trace of disease was discernible. I concluded that it was an old bird that had died solely from natural failure of the life-energy.

But how wonderful, how almost incredible, that the healthy vigour and joy of life should have continued in this individual bird down to within so short a period of the end; that it should have been not only strong enough to find its food, but to rush and wheel about for long intervals in purely sportive exercises, when the brief twilight of decline and final extinction were so near! It becomes credible--we can even believe that most of the individuals that cease to exist only when the vital fire has burnt itself out, fall on death in this swift, easy manner--when we recall the fact that even in the life-history of men such a thing

BIRDS IN A VILLAGE 147

is not unknown. Probably there is not one among my readers who will not be able to recall some such incident in his own circle--the case of someone who lived, perhaps, long past the term usually allotted to man, and who finally passed away without a struggle, without a pang, so that those who were with him found it hard to believe that the spirit had indeed gone. In such cases, the subject has invariably been healthy, although it is hard to believe that, in the conditions we exist in, any man

can have the perfect health that all wild creatures enjoy.

X

AFTER my long talk with the bird-catcher on June 24, and two more talks equally long on the two following days, I found that something of the charm the common had had for me was gone. It was not quite the same as formerly; even the sunshine had a something of conscious sadness in it which was like a shadow. Those merry little brown twitterers that frequently shot across the sky, looking small as insects in the wide blue expanse, and ever and anon dropped swiftly down like showers of aerolites, to lose themselves in the grass and herbage, or perch singing on the topmost dead twigs of a bush, now existed in constant imminent danger--not of that quick merciful destruction which Nature has for her weaklings, and for all that fail to reach her high standard; but of a worse fate, the prison life which is not Nature's ordinance, but one of the cunning larger Ape's abhorred inventions. In—

stead of taking my usual long strolls about the common I loitered once more in the village lanes and had my reward.

On the morning of June 27 I was out sauntering very indolently, thinking of nothing at all; for it was a surpassingly brilliant day, and the sunshine produced the effect of a warm, lucent, buoyant fluid, in which I seemed to float rather than walk--a celestial water, which, like the more ponderable and common sort, may sometimes be both felt and seen. The sensation of feeling it is somewhat similar to that experienced by a bather standing breast-deep in a dear, green, warm tropical sea, so charged with salt that it lifts him up; but to distinguish it with the eye, you must look away to a distance of some yards in an open unshaded place, when it will become visible as fine glinting lines, quivering and serpentining upwards, fountain-wise, from the surface. All at once I was startled by hearing the loud importunate hunger-call of a young cuckoo quite close to me. Moving softly up to the low hedge and peering over, I saw the bird perched on a long cross-stick, which had been put up in a cottage garden to hang clothes on; he was not

---

more than three to four yards from me, a fine young cuckoo in perfect plumage, his barred under-surface facing me. Although seeing me as plainly as I saw him, he exhibited no fear, and did not stir. Why should he, since I had not come there to feed him, and, to his inexperienced avian mind, was only one of the huge terrestrial creatures of various forms, with horns and manes on their heads, that move heavily about in roads and pastures, and are nothing to birds? But his foster parent, a hedge-sparrow, was suspicious, and kept at some distance with food in her bill; then excited by his imperative note, she flitted shyly to him, and deposited a minute caterpillar in his great gaping yellow mouth. It was like dropping a bun into the monstrous mouth of the hippopotamus of the Zoological Gardens. But the hedge-sparrow was off and back again with a second morsel in a very few moments; and again and again she darted away in quest of food and returned successful, while the lazy, beautiful giant sat sunning himself on his cross-stick and hungrily cried for more.

This is one of those exceptional sights in nature which, however often seen, never become alto-

---

gether familiar, never fail to re-excite the old feelings of wonder and admiration which were experienced on first witnessing them. I can safely say, I think, that no man has observed so many parasitical young birds (individuals) being fed by their foster-parents as myself, yet the interest such a sight inspired in me is just as fresh now as in boyhood. And probably in no parasitical species does the strangeness of the spectacle strike the mind so sharply as in this British bird, since the differences in size and colouring between the foster-parent and its false offspring are so much greater in its case. Here nature's unnaturalness in such an instinct--a close union of the beautiful and the monstrous--is seen in its extreme form. The hawk-like figure and markings of the cuckoo serve only to accentuate the disparity, which is perhaps greatest when the parent is the hedge-sparrow--so plainly-coloured a bird, so shy and secretive in its habits. One never ceases to be amazed at the blindness of the parental instinct in so intelligent a creature as a bird in a case of this kind. Some idea of how blind it is may be formed by imagining a case in widely separated types of our own species, which would be a

152 BIRDS IN TOWN AND VILLAGE

parallel to that of the cuckoo and hedge-sparrow. Let us imagine that some malicious Arabian Night's genius had snatched up the infant male child of a Scandinavian couple--the largest of their nation; and flying away to Africa with it, to the heart of the great Aruwhimi forest had laid it on the breast of a little coffee-coloured, woolly-headed, spindle-shanked, pot-bellied, pigmy mother, taking away at the same time her own newly-born babe; that she had tenderly nursed the substituted child, and reared and protected it, ministering, according to her lights, to all its huge wants, until he had come to the fullness of his stature, yet never suspected, that the magnificent, ivory-limbed giant, with flowing

yellow locks and cerulean eyes, was not the child of her own womb.

XI

BRIGHT and genial were all the last days of June, when I loitered in the lanes before the unwished day of my return to London. During this quiet, pleasant time the greenfinch was perhaps more to me than any other songster. In the village itself, with the adjacent lanes and orchards, this pretty, seldom-silent bird was the most common species. The village was his metropolis, just as London is ours--and the sparrow's; its lanes were his streets, its hedges and elm trees his cottage rows and tall stately mansions and public buildings. . We frequently find the predominance of one species somewhat wearisome. Speaking for myself, there are songsters that are best appreciated when they are limited in numbers and keep their distance, but of the familiar, unambitious strains of swallow, robin, and wren I never tire, nor, during these days, could I have too much of the greenfinch, low as he ranks among British

melodists. Tastes differ; that is a point on which we are all agreed, and every one of us, even the humblest, is permitted to have his own preferences. Still, after re-reading Wordsworth's lines to "The Green Linnet," it is curious, to say the least of it, to turn to some prosewriter--an authority on birds, perhaps--to find that this species, whose music so charmed the poet, has for its song a monotonous croak, which it repeats at short intervals for hours without the slightest variation--a dismal sound which harmonizes with no other sound in nature, and suggests nothing but heat and weariness, and is of all natural sounds the most irritating. To this writer, then--and there are others to keep him in countenance--the greenfinch as a vocalist ranks lower than the lowest. One can only wonder (and smile) at such extreme divergences. To my mind all natural sounds have, in some measure an exhilarating effect, and I cannot get rid of the notion that so it should be with every one of us; and when some particular sound, or series of sounds, that has more than this common character, and is distinctly pleasing, is spoken of as nothing but disagreeable, irritating, and the rest

of it, I am inclined to think that there is something wrong with the person who thus describes it; that he is not exactly as nature would have had him, but that either during his independent life, or before it at some period of his prenatal existence, something must have happened to distune him. All this, I freely confess, may be nothing but fancy. In any case, the subject need not keep us longer from the greenfinch--that is to say, *my* greenfinch not another man's.

From morning until evening all around and about the cottage, and out of doors whithersoever I bent my steps, from the masses of deep green foliage, sounded the perpetual airy prattle of these delightful birds. One had the idea that the concealed vocalists were continually meeting each other at little social gatherings, where they exchanged pretty loving greetings, and indulged in a leafy gossip, interspersed with occasional fragments of music, vocal and instrumental; now a long trill--a trilling, a tinkling, a sweeping of one minute finger-tip over metal strings as fine as gossamer threads--describe it how you will, you cannot describe it; then the long, low, inflected scream, like a lark's throat-note drawn out and

inflected; little chirps and chirruping exclamations and remarks, and a soft warbled note three or four or more times repeated, and sometimes, the singer fluttering up out of the foliage and hovering in the air, displaying his green and yellow plumage while emitting these lovely notes; and again the trill, trill answering trill in different keys; and again the music scream, as if some unsubstantial being, fairy or woodnymph had screamed somewhere in her green hiding-place. In London one frequently hears, especially in the spring, half-a-dozen sparrows just met together in a garden tree, or among the ivy or creeper on a wall, burst out suddenly into a confused rapturous chorus of chirruping sounds, mingled with others of a finer quality, liquid and ringing. At such times one is vexed to think that there are writers on birds who invariably speak of the sparrow as a tuneless creature, a harsh chirper, and nothing more. It strikes one that such writers either wilfully abuse or are ignorant of the right meaning of words, so wild and glad in character are these concerts of town sparrows, and so refreshing to the tired and noise-vexed brain! But now when I listened to

---

BIRDS IN A VILLAGE 157

the greenfinches in the village elms and hedgerows, if by chance a few sparrows burst out in loud gratulatory notes, the sounds they emitted appeared coarse, and I wished the chirrupers away. But with the true and brilliant songsters it seemed to me that the rippling greenfinch music was always in harmony, forming as it were a kind of airy, subdued accompaniment to their loud and ringing tones.

I had had my nightingale days, my cuckoo and blackbird and tree-pipit days, with others too numerous to mention, and now I was having my greenfinch days; and these were the last.

One morning in July I was in my sitting-room, when in the hedge on the other side of the lane, just opposite my window, a small brown bird warbled a few rich notes, the prelude to his song. I went and stood by the open window, intently listening, when it sang again, but only a phrase or two. But I listened still, confidently expecting more; for although it was now long past its singing season, that splendid sunshine would compel it to express its gladness. Then, just when a fresh burst of music came, it was disturbed by another sound close by--a human voice, also sing-

---

ing. On the other side of the hedge in which the bird sat concealed was a cottage garden, and there on a swing fastened to a pair of apple trees, a girl about eleven years old sat lazily swinging herself. Once or twice after she began singing the nightingale broke out again, and then at last he became silent altogether, his voice overpowered by hers. Girl and bird were not five yards apart. It greatly surprised me to hear her singing, for it was eleven o'clock, when all the village children were away at the National School, a time of day when, so far as human sounds were concerned, there reigned an almost unbroken silence. But very soon I recalled the fact that this was a very lazy child, and concluded that she had coaxed her mother into sending an excuse for keeping her at home, and so had kept her liberty on this beautiful morning. About two minutes' walk from the cottage, at the side of the crooked road running through the village, there was a group of ancient pollarded elm trees with huge, hollow trunks, and behind them an open space, a pleasant green slope, where some of the village children used to go every day to play on the grass. Here I used to see this girl lying in the sun, her dark

---

chestnut hair loosed and scattered on the sward, her arms stretched out, her eyes nearly closed, basking in the sun, as happy as some heat-loving wild animal. No, it was not strange that she had not gone to school with the others when her disposition was remembered, but most strange to hear a voice of such quality in a spot where nature was rich and lovely, and only man was, if not vile, at all events singularly wanting in the finer human qualities.

Looking out from the open window across the low hedge-top, I could see her as she alternately rose and fell with slow, indolent motion, now waist-high above the green dividing wall, then only her brown head visible resting against the rope just where her hand had grasped it. And as she swayed herself to and fro she sang that simple melody--probably some child's hymn which she had been taught at the Sunday-school; but it was a very long hymn, or else she repeated the same few stanzas many times, and after each there was a brief pause, and then the voice that seemed to fall and rise with the motion went on as before. I could have stood there for an hour--nay, for hours--listening to it, so fresh and

so pure was the clear young voice, which had no earthly trouble in it, and no passion, and was in this like the melody of the birds of which I had lately heard so much; and with it all that tenderness and depth which is not theirs, but is human only and of the soul.

It struck me as a singular coincidence--and to a mind of so primitive a type as the writer's there is more in the fact that the word implies--that, just as I had quitted London, to seek for just such a spot as I so speedily found, with the passionately exclaimed words of a young London girl ringing in my ears, so now I went back with this village girl's melody sounding and following me no less clearly and insistently. For it was not merely remembered, as we remember most things, but vividly and often reproduced, together with the various melodies of the birds I had listened to; a greater and principal voice in that choir, yet in no wise lessening their first value, nor ever out of harmony with them.

EXOTIC BIRDS FOR BRITAIN

THERE are countries with a less fertile soil and a worse climate than ours, yet richer in bird life. Nevertheless, England is not poor; the species are not few in number, and some are extremely abundant. Unfortunately many of the finer kinds have been too much sought after; persecuted first for their beauty, then for their rarity, until now we are threatened with their total destruction. As these kinds become unobtainable, those which stand next in the order of beauty and rarity are persecuted in their turn; and in a country as densely populated as ours, where birds cannot hide themselves from human eyes, such persecution must eventually cause their extinction. Meanwhile the bird population does not decrease. Every place in nature, like every property in

Chancery, has more than one claimant to it--sometimes the claimants are many--and so long as the dispute lasts all live out of the estate. For there are always two or more species subsisting on the same kind of food, possessing similar habits, and frequenting the same localities. It is consequently impossible for man to exterminate any one species without indirectly benefiting some other species, which attracts him in a less degree, or not at all. This is unfortunate, for as the bright kinds, or those we esteem most, diminish in numbers the less interesting kinds multiply, and we lose much of the pleasure which bird life is fitted to give us. When we visit woods, or other places to which birds chiefly resort, in districts uninhabited by man, or where he pays little or no attention to the feathered creatures, the variety of the bird life encountered affords a new and peculiar delight. There is a constant succession of new forms and new voices; in a single day as many species may be met with as one would find in England by searching diligently for a whole year.

And yet this may happen in a district possessing no more species than England boasts; and the actual number of individuals may be even less than with us. In sparrows, for instance, of the one common species, we are exceedingly rich; but in bird life generally, in variety of birds, especially in those of graceful forms and beautiful plumage, we have been growing poorer for the last fifty years, and have now come to so low a state that it becomes us to inquire whether it is not in our power to better ourselves. It is an old familiar truth--a truism--that it is easier to destroy than to restore or build up; nevertheless, some comfort is to be got from the reflection that in this matter we have up till now been working against Nature. She loves not to bring forth food where there are none to thrive on it; and when our unconsidered action had made these gaps, when, despising her gifts or abusing them, we had destroyed or driven out her finer kinds, she fell back on her lowlier kinds--her reserve of coarser, more generalized species--and gave them increase, and bestowed the vacant places which we had created on them. What she has done she will undo, or assist us in undoing; for we should be going back to her methods, and should have her with and not againstus. Much might yet be

---

done to restore the balance among our native species. Not by legislation, albeit all laws restraining the wholesale destruction of bird life are welcome. On this subject the Honourable Auberon Herbert has said, and his words are golden: "For myself, legislation or no legislation, I would turn to the friends of animals in this country, and say, 'If you wish that the friendship between man and animals should become a better and truer thing than it is at present, you must make it so by countless individual efforts, by making thousands of centres of personal influence.'"

The subject is a large one. In this paper the question of the introduction of exotic birds will be chiefly considered. Birds have been blown by the winds of chance over the whole globe, and have found rest for their feet. That a large number of species, suited to the conditions of this country, exist scattered about the world is not to be doubted, and by introducing a few of these we might accelerate the change so greatly to be desired. At present a very considerable amount of energy is spent in hunting down the small contingents of rare species that once inhabited our

---

EXOTIC BIRDS FOR BRITAIN 165

islands, and still resort annually to its shores, persistently endeavouring to re-establish their colonies. A less amount of labour and expense would serve to introduce a few foreign species each year, and the reward would be greater, and would not make us ashamed. We have generously given our own wild animals to other countries; and from time to time we receive cheering reports of an abundant increase in at least two of our exportations--to wit, the rabbit and the sparrow. We are surely entitled to some return. Dead animals, however rich their pelt or bright their plumage may be, are not a fair equivalent. Dead things are too much with us. London has become a mart for this kind of merchandise for the whole of Europe, and the traffic is not without a reflex effect on us; for life in the inferior animals has come or is coming to be merely a thing to be lightly taken by human hands, in order that its dropped garment may be sold for filthy lucre. There are warehouses in this city where it is possible for a person to walk ankle-deep--literally to wade--in bright-plumaged bird-skins, and see them piled shoulder-high on either side of him--a sight to make the angels weep I

---

166 BIRDS IN TOWN AND VILLAGE

Not the angel called woman. It is not that she is naturally more cruel than man; bleeding wounds and suffering in all its forms, even the sigh of a burdened heart, appeal to her quick sympathies, and draw the ready tears; but her imagination helps her less. The appeal must in most cases be direct and through the medium of her senses, else it is not seen and not heard. If she loves the ornament of a gay-winged bird, and is able to wear it with a light heart, it is because it calls up no mournful image to her mind; no little tragedy enacted in some far-off wilderness, of the swift child of the air fallen and bleeding out its bright life, and its callow nestlings, orphaned of the breast that warmed them, dying of hunger in the tree. We know, at all events, that out of a female population of many millions in this country, so far only ten women, possibly fifteen, have been found to raise their voices--raised so often and so loudly on other questions--to protest against the barbarous and abhorrent fashion of wearing slain birds as ornaments. The degrading business of supplying the demand for this kind of feminine adornment must doubtless continue to flourish in our midst, commerce not

---

being compatible with morality, but the material comes from other lands, unblessed as yet with Wild Bird Protection Acts, and "individual efforts, and thousands of centres of personal influence"; it comes mainly from the tropics, where men have brutish minds and birds a brilliant plumage. This trade, therefore, does not greatly affect the question of our native bird life, and the consideration of the means, which may be within our reach, of making it more to us than it now is. Some species from warm and even hot climates have been found to thrive well in England, breeding in the open air; as, for instance, the black and the black-necked swans, the Egyptian goose, the mandarin and summer ducks, and others too numerous to mention. But these birds are semi-domestic, and are usually kept in enclosures, and that they can stand the climate and propagate when thus protected from competition is not strange; for we know that several of our hardy domestic birds--the fowl, pea-fowl, Guinea-fowl, and Muscovy duck--are tropical in their origin. Furthermore, they are all comparatively large, and if they ever become feral in England, it will not be for many years to come.

---

168 BIRDS IN TOWN AND VILLAGE

That these large kinds thrive so well with us is an encouraging fact; but the question that concerns us at present is the feasibility of importing birds of the grove, chiefly of the passerine order, and sending them forth to give a greater variety and richness to our bird life. To go with such an object to tropical countries would only be to court failure. Nature's highest types, surpassing all others in exquisite beauty of form, brilliant colouring, and perfect melody, can never be known to our woods and groves. These rarest avian gems may not be removed from their setting, and to those who desire to know them in their unimaginable lustre, it will always be necessary to cross oceans and penetrate into remote wildernesses. We must go rather to regions where the conditions of life are hard, where winters are long and often severe, where Nature is not generous in the matter of food, and the mouths are many, and the competition great. Nor even from such regions could we take any strictly migratory species with any prospect of success. Still, limiting ourselves to the resident, and consequently to the hardiest kinds, and to those possessing only a partial migration, it is surprising to find how

EXOTIC BIRDS FOR BRITAIN 169

many there are to choose from, how many are charming melodists, and how many have the bright tints in which our native species are so sadly lacking. The field from which the supply can be drawn is very extensive, and includes the continent of Europe, the countries of North Asia, a large portion of North America and Antarctic America, or South Chili and Patagonia. It would not be going too far to say that for every English species, inhabiting the garden, wood, field, stream, or waste, at least half a dozen resident species, with similar habits, might be obtained from the countries mentioned which would be superior to our own in melody (the nightingale and lark excepted), bright plumage, grace of form, or some other attractive quality. The question then arises; What reason is there for believing that these exotics, imported necessarily in small numbers, would succeed in winning a footing in our country, and become a permanent addition to its avifauna? For it has been admitted that our species are not few, in spite of the losses that have been suffered, and that the bird population does not diminish, however much its character may have altered and

deteriorated from the aesthetic point of view, and probably also from the utilitarian. There are no vacant places. Thus, the streams are fished by herons, grebes, and kingfishers, while the rushy margins are worked by coots and gallinules, and, above the surface, reed and sedge-warblers, with other kinds, inhabit the reed-beds. The decaying forest tree is the province of the woodpecker, of which there are three kinds; and the trunks and branches of all trees, healthy or decaying, are quartered by the small creeper, that leaves no crevice unexplored in its search for minute insects and their eggs. He is assisted by the nuthatch; and in summer the wryneck comes (if he still lives), and deftly picks up the little active ants that are always wildly careering over the boles. The foliage is gleaned by warblers and others; and not even the highest terminal twigs are left unexamined by tits and their fellow-seekers after little things. Thrushes seek for worms in moist grounds about the woods; starlings and rooks go to the pasture lands; the lark and his relations keep to the cultivated fields; and there also dwells the larger partridge. Waste and stony grounds are occupied by the chats, and even on the barren

mountain summits the ptarmigan gets his living. Wagtails run on the clean margins of streams; and littoral birds of many kinds are in possession of the entire sea-coast. Thus, the whole ground appears to be already sufficiently occupied, the habitats of distinct species overlapping each other like the scales on a fish. And when we have enumerated all these, we find that scores of others have been left out. The important flycatcher; the wren, Nature's diligent little housekeeper, that leaves no dusty corner uncleaned; and the pigeons, that have a purely vegetable diet. The woods and thickets are also ranged by jays, cuckoos, owls, hawks, magpies, butcher-birds--Nature's gamekeepers, with a licence to kill, which, after the manner of gamekeepers, they exercise somewhat indiscriminately. Above the earth, the air is peopled by swifts and swallows in the daytime, and by goatsuckers at night. And, as if all these were not enough, the finches are found scattered everywhere, from the most secluded spot in nature to the noisy public thoroughfare, and are eaters of most things, from flinty seed to softest caterpillar. This being the state of things, one might imagine that experience and

172 BIRDS IN TOWN AND VILLAGE

observation are scarcely needed to prove to us that the exotic, strange to the conditions, and where its finest instincts would perhaps be at fault, would have no chance of surviving. Nevertheless, odd as it may seem, the small stock of facts bearing on the subject which we possess point to a contrary conclusion. It might have been assumed, for instance, that the red-legged partridge would never have established itself with us, where the ground was already fully occupied by a native species, which possessed the additional advantage of a more perfect protective colouring. Yet, in spite of being thus handicapped, the stranger has conquered a place, and has spread throughout the greater part of England. Even more remarkable is the case of the pheasant, with its rich plumage, a native of a hot region; yet our cold, wet climate and its unmodified bright colours have not been fatal to it, and practically it is one of our wild birds. The large capercailzie has also been successfully introduced from Norway. Small birds would probably become naturalized much more readily than large ones; they are volatile, and can more quickly find suitable feeding-ground, and safe

EXOTIC BIRDS FOR BRITAIN 173

roosting and nesting places; their food is also more abundant and easily found; their small size, which renders them inconspicuous, gives them safety; and, finally, they are very much more adaptive than large birds.

It is not at all probable that the red-legged partridge will ever drive out our own bird, a contingency which some have feared. That would be a misfortune, for we do not wish to change one bird for another, or to lose any species we now possess, but to have a greater variety. We are better off with two partridges than we were with one, even if the invader does not afford such good sport nor such delicate eating. They exist side by side, and compete with each other; but such competition is not necessarily destructive to either. On the contrary, it acts and re-acts healthily and to the improvement of both. It is a fact that in small islands, very far removed from the mainland, where the animals have been exempt from all foreign competition--that is, from the competition of casual colonists--when it does come it proves, in many cases, fatal to them. Fortunately, this country's large size and nearness to the mainland has prevented any such

174 BIRDS IN TOWN AND VILLAGE

fatal crystallization of its organisms as we see in islands like St. Helena. That any English species would be exterminated by foreign competition is extremely unlikely; whether we introduce exotic birds or not, the only losses we shall have to deplore in the future will, like those of the past, be directly due to our own insensate action in slaying every rare and beautiful thing with powder and shot. From the introduction of exotic species nothing is to be feared, but much to be hoped.

There is another point which should not be overlooked. It has after all become a mere fiction to say that *all* places are occupied. Nature's nice order has been destroyed, and her kingdom thrown into the utmost confusion; our action tends to maintain the disorderly condition, while she is perpetually working against us to re-establish order. When she multiplies some common, little-regarded species to occupy a space left vacant by an artificially exterminated kind, the species called in as a mere stop-gap, as it were, is one not specially adapted in structure and instincts to a particular mode of life, and consequently cannot fully and effectually occupy the

HERON.
. . . the streams are fished by herons.'

EXOTIC BIRDS FOR BRITAIN 175

ground into which it has been permitted to enter. To speak in metaphor, it enters merely as a caretaker or ignorant and improvident steward in the absence of the rightful owner. Again, some of our ornamental species, which are fast diminishing, are fitted from their peculiar structure and life habits to occupy places in nature which no other kinds, however plastic they may be, can even partially fill. The wryneck and the woodpecker may be mentioned; and a still better instance is afforded by the small, gem-like kingfisher

--the only British bird which can properly be described as gem-like. When the goldfinch goes

--and we know that he is going rapidly--other coarser fringilline birds, without the melody, brightness, and charm of the goldfinch--sparrow and bunting--come in, and in some rough fashion supply its place; but when the kingfisher disappears an important place is left absolutely vacant, for in this case there is no coarser bird of homely plumage with the fishing instinct to seize upon it. Here, then, is an excellent opportunity for an experiment. In the temperate regions of the earth there are many fine kingfishers to select from; some are resident in countries colder than

---

176 BIRDS IN TOWN AND VILLAGE

England, and are consequently very hardy; and in some cases the rivers and streams they frequent are exceedingly poor in fish. Some of them are very beautiful, and they vary in size from birds no larger than a sparrow to others as large as a pigeon.

Anglers might raise the cry that they require all the finny inhabitants of our waters for their own sport. It is scarcely necessary to go as deeply into the subject as mathematical-minded Mudie did to show that Nature's lavishness in the production of life would make such a contention unreasonable. He demonstrated that if all the fishes hatched were to live their full term, in twenty-four years their production power would convert into fish (two hundred to the solid foot) as much matter as there is contained in the whole solar system--sun, planets, and satellites! An "abundantly startling" result, as he says. To be well within the mark, ninety-nine out of every hundred fishes hatched must somehow perish during that stage when they are nothing but suitable morsels for the kingfisher, to be swallowed entire; and a portion of all this wasted food might very well go to sustain a few species, which would be beautiful ornaments of the waterside, and a perpetual delight to all lovers of rural nature, including anglers. It may be remarked in passing, that the waste of food, in the present disorganized state of nature, is not only in our streams.

The introduction of one or more of these lovely foreign kingfishers would not certainly have the effect of hastening the decline of our native species; but indirectly it might bring about a contrary result--a subject to be touched on at the end of this paper. Practical naturalists may say that kingfishers would be far more difficult to procure than other birds, and that it would be almost impossible to convey them to England. That is a question it would be premature to discuss now; but if the attempt should ever be made, the difficulties would not perhaps be found insuperable. In all countries one hears of certain species of birds that they invariably die in captivity; but when the matter is closely looked into, one usually finds that improper treatment and not loss of liberty is the cause of death. Unquestionably it would be much more difficult to keep a kingfisher alive and healthy during a long sea-voyage than a common seed-eating bird; but the

same may be said of woodpeckers, cuckoos, warblers, and, in fact, of any species that subsists in a state of nature on a particular kind of animal food. Still, when we find that even the excessively volatile humming-bird, which subsists on the minutest insects and the nectar of flowers, and seems to require unlimited space for the exercise of its energies, can be successfully kept confined for long periods and conveyed to distant countries, one would imagine that it would be hard to set a limit to what might be done in this direction. We do not want hard-billed birds only. We require, in the first place, variety; and, secondly, that every species introduced, when not of type unlike any native kind, as in the case of the pheasant, shall be superior in beauty, melody, or some other quality, to its British representative, or to the species which comes nearest to it in structure and habits. Thus, suppose that the introduction of a pigeon should be desired. We know that in all temperate regions, these birds vary as little in colour and markings as they do in form; but in the vocal powers of different species there is great diversity; and the main objects would therefore be to secure a bird which

---

would be an improvement in this respect on the native kinds. There are doves belonging to the same genus as stockdove and wood-pigeon, that have exceedingly good voices, in which the peculiar mournful dove-melody has reached its highest perfection--weird and passionate strains, surging and ebbing, and startling the hearer with their mysterious resemblance to human tones. Or a Zenaida might be preferred for its tender lament, so wild and exquisitely modulated, like sobs etherealized and set to music, and passing away in sigh-like sounds that seem to mimic the aerial voices of the wind.

When considering the character of our bird population with a view to its improvement, one cannot but think much, and with a feeling almost of dismay, of the excessive abundance of the sparrow. A systematic persecution of this bird would probably only serve to make matters worse, since its continued increase is not the cause but an effect of a corresponding decrease in other more useful and attractive species; and if Nature is to have her way at all there must be birds; and besides, no bird-lover has any wish at see such a thing attempted. The sparrow has

---

his good points, if we are to judge him as we find him, without allowing what the Australians and Americans say of him to prejudice our minds. Possibly in those distant countries he may be altogether bad, resembling, in this respect, some of the emigrants of our species, who, when they go abroad, leave their whole stock of morality at home. Even with us Miss Ormerod is exceedingly bitter against him, and desires nothing less than his complete extirpation; but it is possible that this lady's zeal may not be according to knowledge, that she may not know a sparrow quite so well as she knows a fly. At all events, the ornithologist finds it hard to believe that so bad an insect-catcher is really causing the extinction of any exclusively insectivorous species. On her own very high authority we know that the insect supply is not diminishing, that the injurious kinds alone are able to inflict an annual loss equal to PS10,000,000 on the British farmer. To put aside this controversial matter, the sparrow with all his faults is a pleasant merry little fellow; in many towns he is the sole representative of wild bird life, and is therefore a great deal to us--especially in the metropolis, in which he most

---

EXOTIC BIRDS FOR BRITAIN 181

abounds, and where at every quiet interval his blithe chirruping comes to us like a sound of subdued and happy laughter. In London itself this merriment of Nature never irritates; it is so much finer and more aerial in character than the gross jarring noises of the street, that it is a relief to listen to it, and it is like melody. In the quiet suburbs it sounds much louder and without intermission. And going further afield, in woods, gardens, hedges, hamlets, towns--everywhere there is the same running, rippling sound of the omnipresent sparrow, and it becomes monotonous at last. We have too much of the sparrow. But we are to blame for that. He is the unskilled worker that Nature has called in to do the work of skilled hands, which we have foolishly turned away. He is willing enough to take it all on himself; his energy is great; he bungles away without ceasing; and being one of a joyous temperament, he whistles and sings in his tuneless fashion at his work, until, like the grasshopper of Ecclesiastes, he becomes a burden. For how tiring are the sight and sound of grasshoppers when one journeys many miles and sees them incessantly rising like a sounding cloud before his

---

horse, and hears their shrill notes all day from the wayside! Yet how pleasant to listen to their minstrelsy in the green summer foliage, where they are not too abundant! We can have too much of anything, however charming it may be in itself. Those who live where scores of humming-birds are perpetually dancing about the garden flowers find that the eye grows weary of seeing the daintiest forms and brightest colours and liveliest motions that birds exhibit. We are told that Edward the Confessor grew so sick of the incessant singing of nightingales in the forest of Havering-at-Bower that he prayed to Heaven to silence their music; whereupon the birds promptly took their departure, and returned no more to that forest until after the king's death. The sparrow is not so sensitive as the legendary nightingales, and is not to be got rid of in this easy manner. He is amenable only to a rougher kind of persuasion; and it would be impossible to devise a more effectual method of lessening his predominance than that which Nature teaches--namely to subject him to the competition of other and better species. He is well equipped for the struggle--hardy, pugnacious, numerous, and in

possession. He would not be in possession and so predominant if he had not these qualities, and great pliability of instinct and readiness to seize on vacant places. Nevertheless, even with the sturdy sparrow a very small thing might turn the scale, particularly if we were standing by and putting a little artificial pressure on one side of the balance; for it must be borne in mind that the very extent and diversity of the ground he occupies is a proof that he does not occupy it effectually, and that his position is not too strong to be shaken. It is not probable that our action in assisting one side against the other would go far in its results; still, a little might be done. There are gardens and grounds in the suburbs of London where sparrows are not abundant, and are shyer than the birds of other species, and this result has been brought about by means of a little judicious persecution. Shooting is a bad plan, even with an air-gun;

its effects are seen by all the birds, for they see more from their green hiding-places than we imagine, and it creates a general alarm among them. Those who wish to give the other birds a chance will only defeat their own object by shooting the sparrows. A

---

much better plan for those who are able to practise it prudently is to take their nests, which are more exposed to sight than those of other birds; but they should be taken after the full complement of eggs have been laid, and only at night, so that other birds shall not witness the robbery and fear for their own treasures. Mr. Henry George, in that book of his which has been the delight of so many millions of rational souls, advocates the destruction of all sharks and other large rapacious fishes, after which, he says, the ocean can be stocked with salmon, which would secure an unlimited supply of good wholesome food for the human race. No such high-handed measures are advocated here with regard to the sparrow. Knowledge of nature makes us conservative. It is so very easy to say, "Kill the sparrow, or shark, or magpie, or whatever it is, and then everything will be right." But there are more things in nature than are dreamt of in the philosophy of the class of reformers represented by the gamekeeper, and the gamekeeper's master, and Miss Ormerod, and Mr. Henry George. Let him by all means kill the sharks, but he will not conquer Nature in that way: she

---

will make more sharks out of something else--possibly out of the very salmon on which he proposes to regale his hungry disciples. To go into details is not the present writer's purpose; and to finish with this part of the subject, it is sufficient to add that in the very wide and varied field occupied by the sparrow, in that rough, ineffectual manner possible to a species having no special and highly perfected feeding instincts, there is room for the introduction of scores of competitors, every one of which should be better adapted than the sparrow to find a subsistence at that point or that particular part of the field where the two would come into rivalry; and every species introduced should also possess some quality which would make it, from the aesthetic point of view, a valuable addition to our bird life. This would be no war of violence, and no contravention of Nature's ordinances, but, on the contrary, a return to her safe, healthy, and far-reaching methods.

There is one objection some may make to the scheme suggested here which must be noticed. It may be said that even if exotic species able to thrive in our country were introduced there

---

would be no result; for these strangers to our groves would all eventually meet with the same fate as our rarer species and casual visitors--that is to say, they would be shot. There is no doubt that the amateur naturalist has been a curse to this country for the last half century, that it is owing to the "cupidity of the cabinet" as old Robert Mudie has it--that many of our finer species are exceedingly rare, while others are disappearing altogether. But it is surely not too soon to look for a change for the better in this direction. Half a century ago, when the few remaining great bustards in this country were being done to death, it was suddenly remembered by naturalists that in their eagerness to possess examples of the bird (in the skin) they had neglected to make themselves acquainted with its customs when alive. Its habits were hardly better known than those of the dodo and solitaire. The reflection came too late, in so far as the habits of the bird in this country are concerned; but unhappily the lesson was not then taken to heart, and other fine species have since gone the way of the great bustard. But now that we have so clearly seen the disastrous effects of this method

---

of "studying ornithology," which is not in harmony with our humane civilization, it is to be hoped that a better method will be adopted --that "finer way" which Thoreau found and put aside his fowling-piece to practise. There can be no doubt that the desire for such an improvement is now becoming very general, that a kindlier feeling for animal, and especially bird life is growing up among us, and there are signs that it is even beginning to have some appreciable effect. The fashion of wearing birds is regarded by most men with pain and reprobation; and it is possible that before long it will be thought that there is not much difference between the action of the woman who buys tanagers and humming-birds to adorn her person, and that of the man who kills the bittern, hoopoe, waxwing, golden oriole, and Dartford-warbler to enrich his private collection.

A few words on the latest attempt which has been made to naturalize an exotic bird in England will not seem out of place here. About eight years ago a gentleman in Essex introduced the rufous tinamou--a handsome game bird, nearly as large as a fowl--into his estate. Up till the

present time, or till quite recently these birds have bred every year, and at one time they had increased considerably and scattered about the neighbourhood. When it began to increase, the neighbouring proprietors and sportsmen generally were asked not to shoot it, but to give it a chance, and there is reason to believe that they have helped to protect it, and have taken a great interest in the experiment. Whatever the ultimate result may be, the partial success attained during these few years is decidedly encouraging, and that for more reasons than one. In the first place, the bird was badly chosen for such an experiment. It belongs to the pampas of La Plata, to which it is restricted, and where it enjoys a dry, bright climate, and lives concealed in the tall close-growing indigenous grasses. The conditions of its habitat are therefore widely different from those of Essex, or of any part of England; and, besides, it has a peculiar organisation, for it happens to be one of those animals of ancient types of which a few species still survive in South America. That so unpromising a subject as this large archaic tinamou should be able to maintain its existence in this country, even for a very few

years, encourages one to believe that with better-chosen species, more highly organized, and with more pliant habits, such as the hazel hen of Europe for a game bird, success would be almost certain.

Another circumstance connected with the attempted introduction of this unsuitable bird, even of more promise than the mere fact of the partial success achieved, is the greatest interest the experiment has excited, not only among naturalists throughout the country, but also among landlords and sportsmen down in Essex, where the bird was not regarded merely as fair game to be bagged, or as a curiosity to be shot for the collector's cabinet, but was allowed to fight its own fight without counting man among its enemies. And it is to be expected that the same self-restraint and spirit of fairness and intelligent desire to see a favourable result would be shown everywhere if exotic species were to be largely introduced, and breeding centres established in suitable places throughout the country. When it once became known that individuals were doing this thing, giving their time and best efforts and at considerable expense not for their own selfish gratification, but for the

general good, and to make the country more delightful to all lovers of rural sights and sounds, there would be no opposition, but on the contrary every assistance, since all would wish success to such an enterprise. Even the most enthusiastic collector would refrain from lifting a weapon against the new feathered guests from distant lands; and if by any chance an example of one should get into his hands he would be ashamed to exhibit it.

The addition of new beautiful species to our avifauna would probably not be the only, nor even the principal benefit we should derive from the carrying out of the scheme here suggested. The indirect effect of the knowledge all would possess that such an experiment was being conducted, and that its chief object was to repair the damage that has been done, would be wholly beneficial since it would enhance the value in our eyes of our remaining native rare and beautiful species. A large number of our finer birds are annually shot by those who know that they are doing a great wrong--that if their transgression is not punishable by law it is really not less grave than that of the person who maliciously barks a

---

EXOTIC BIRDS FOR BRITAIN 191

shade tree in a park or public garden--but who excuse their action by saying that such birds must eventually get shot, and that those who first see them might as well have the benefit. The presence of even a small number of exotic species in our woods and groves would no doubt give rise to a better condition of things; it would attract public attention to the subject; for the birds that delight us with their beauty and melody should be for the public, and not for the few barbarians engaged in exterminating them; and the "collector" would find it best to abandon his evil practices when it once began to be generally asked, if we can spare the rare, lovely birds brought hither at great expense from China or Patagonia, can we not also spare our own kingfisher, and the golden oriole, and the hoopoe, that comes to us annually from Africa to breed, but is not permitted to breed, and many other equally beautiful and interesting species?

MOORHENS IN HYDE PARK

THE sparrow, like the poor, we have always with us, and on windy days even the large-sized rook is blown about the murkiness which does duty for sky over London; and on such occasions its coarse, corvine dronings seem not unmusical, nor without something of a tonic effect on our jarred nerves. And here the ordinary Londoner has got to the end of his ornithological list--that is to say, his winter list. He knows nothing about those wind-worn waifs, the "occasional visitors" to the metropolis--the pilgrims to distant Meccas and Medinas that have fallen, overcome by weariness, at the wayside; or have encountered storms in the great aerial sea, and lost compass and reckoning, and have been lured by false lights to perish miserably at the hands of their cruel

enemies. It may be true that gulls are seen on the Serpentine, that woodcocks are flushed in Lincoln's Inn Fields, but the citizen who goes to his office in the morning and returns after the lamps have been lighted, does not see them, and they are nothing in his life. Those who concern themselves to chronicle such incidents might just as well, for all that it matters to him, mistake their species, like that bird-loving but unornithological correspondent of the Times who wrote that he had seen a flock of golden orioles in Kensington Gardens. It turned out that what he had seen were wheatears, or they might draw a little on their imaginations, and tell of sunward-sailing cranes encamped on the dome of St. Paul's Cathedral, flamingoes in the Round Pond, great snowy owls in Westminster Abbey, and an ibis--scarlet, glossy, or sacred, according to fancy--perched on Peabody's statue, at the Royal Exchange.

But his winter does not last for ever. When the bitter months are past, with March that mocks us with its crown of daffodils; when the sun shines, and the rain is soon over; and elms and limes in park and avenue, and unsightly smoke-

---

blackened brushwood in the squares, are dressed once more in tenderest heart-refreshing green, even in London we know that the birds have returned from beyond the sea. Why should they come to us here, when it would seem so much more to their advantage, and more natural for them to keep aloof from our dimmed atmosphere, and the rude sounds of traffic, and the sight of many people going to and fro? Are there no silent green retreats left where the conditions are better suited to their shy and delicate natures? Yet no sooner is the spring come again than the birds are with us. Not always apparent to the eye, but everywhere their irrepressible gladness betrays their proximity; and all London is ringed round with a mist of melody, which presses on us, ambitious of winning its way even to the central heart of our citadel, creeping in, mist-like, along gardens and tree-planted roads, clinging to the greenery of parks and squares, and floating above the dull noises of the town as clouds fleecy and ethereal float above the earth.

Among our spring visitors there is one which is neither aerial in habits, nor a melodist, yet is eminently attractive on account of its graceful

---

form, pretty plumage, and amusing manners; nor must it be omitted as a point in its favour that it is not afraid to make itself very much at home with us in London.* This is the little moor-hen, a bird possessing some strange customs, for which those who are curious about such matters may consult its numerous biographies. Every spring a few individuals of this species make their appearance in Hyde Park, and settle there for the season, in full sight of the fashionable world; for their breeding-place happens to be that minute transcript of nature midway between the Dell and Rotten Row, where a small bed of rushes and aquatic grasses flourishes in the stagnant pool forming the end of the Serpentine. Where they pass the winter--in what Mentone or Madeira of the ralline race--is not known. There is a pretty story, which circulated throughout Europe a little over fifty years ago, of a Polish gentleman, capturing a stork that built its nest on his roof every summer, and putting an iron collar on its neck

*Note that when this was written in 1893, the moor-hen was never known to winter in London; his habits have changed in this respect during the last two decades: he is now a permanent resident.

---

196 BIRDS IN TOWN AND VILLAGE

with the inscription, "Haec Ciconia ex Polonia." The following summer it reappeared with something which shone very brightly on its neck, and when the stork was taken again this was found to be a collar of gold, with which the iron collar had been replaced, and on it were graven the words, "India cum donis remittit ciconian Polonis." No person has yet put an iron collar on the moor-hen to receive gifts in return, or followed its feeble fluttering flight to discover the limits of its migration which is probably no further away than the Kentish marshes and other wet sheltered spots in the south of England; that it leaves the country when it quits the park is not to be believed. Still, it goes with the wave, and with the wave returns; and, like the migratory birds that observe times and seasons, it comes back to its own home--that circumscribed spot of earth and water which forms its little world, and is more to it than all other reedy and willow-shaded pools and streams in England. It is said to be shy in disposition, yet all may see it here, within a few feet of the Row, with so many people continually passing, and so many pausing to watch the pretty birds as they trip about their little plot

---

MOORHEN.
" . . . that quaint, graceful, jetting gait."

MOORHENS IN HYDE PARK 197

of green turf, deftly picking minute insects from the grass and not disdaining crumbs thrown by the children. A dainty thing to look at is that smooth, olive-brown little moor-hen, going about with such freedom and ease in its small dominion, lifting its green legs deliberately, turning its yellow beak and shield this way and that, and displaying the snow-white undertail at every step, as it moves with that quaint, graceful, jetting gait peculiar to the gallinules.

Such a fact as this--and numberless facts just as significant all pointing to the same conclusion, might be adduced--shows at once how utterly erroneous is that often-quoted dictum of Darwin's that birds possess an instinctive or inherited fear of man. These moor-hens fear him not at all; simply because in Hyde Park they are not shot at, and robbed of their eggs or young, nor in any way molested by him. They fear no living thing, except the irrepressible small dog that occasionally bursts into the enclosure, and hunts them with furious barkings to their reedy little refuge. And as with these moor-hens, so it is with all wild birds; they fear and fly from, and suspiciously watch from a safe distance, whatever molests

---

them, and wherever man suspends his hostility towards them they quickly outgrow the suspicion which experience has taught them, or which is traditional among them; for the young and inexperienced imitate the action of the adults they associate with, and learn the suspicious habit from them.

It is also interesting and curious to note that a bird which inhabits two countries, in summer and winter, regulates his habits in accordance with the degree of friendliness or hostility exhibited towards him by the human inhabitants of the respective areas. The bird has in fact two traditions with regard to man's attitude towards him--one for each country. Thus, the field-fare is an exceedingly shy bird in England, but when he returns to the north if his breeding place is in some inhabited district in northern Sweden or Norway he loses all his wildness and builds his nest quite close to the houses. My friend Trevor Battye saw a pair busy making their nest in a small birch within a few yards of the front door of a house he was staying at. "How strange," said he to the man of the house, "to see field-fares making a nest in such a place!"

"Why strange?" said the man in surprise. "Why strange? Because of the boys, always throwing stones at a bird. The nest is so low down, that any boy could put his hand in and take the eggs." "Take the eggs!" cried the man, more astonished than ever. "And throwing stones at a bird! Who ever heard of a boy doing such things!"

Closely related to this error is another error, which is that noise in itself is distressing to birds, and has the effect of driving them away. To all sounds and noises which are not associated with danger to them, birds are absolutely indifferent. The rumbling of vehicles, puffing and shrieking of engines, and braying of brass bands, alarm them less than the slight popping of an air gun, where that modest weapon of destruction is frequently used against them. They have no "nerves" for noise, but the apparition of a small boy silently creeping along the hedge-side, in search of nests or throwing stones, is very terrifying to them. They fear not cattle and horses, however loud the bellowing may be; and if we were to transport and set loose herds of long-necked camelopards, trumpeting elephants, and rhinoceroses of horrible aspect, the little birds

---

200 BIRDS IN TOWN AND VILLAGE

would soon fear them as little as they do the familiar cow. But they greatly fear the small-sized, quiet, unobtrusive, and meek-looking cat. Sparrows and starlings that fly wildly at the shout of a small boy or the bark of a fox-terrier, build their nests under every railway arch; and the incubating bird sits unalarmed amid the iron plates and girders when the express train rushes overhead, so close to her that one would imagine that the thunderous jarring noise would cause the poor thing to drop down dead with terror. To this indifference to the mere harmless racket of civilization we owe it that birds are so numerous around, and even in, London; and that in Kew Gardens, which, on account of its position on the water side, and the numerous railroads surrounding it, is almost as much tortured with noise as Willesden or Clapham Junction, birds are concentrated in thousands. Food is not more abundant there than in other places; yet it would be difficult to find a piece of ground of the same extent in the country proper, where all is silent and there are no human crowds, with so large a bird population. They are more numerous in Kew than elsewhere, in spite of the noise and the

---

MOORHENS IN HYDE PARK 201

people, because they are partially protected there from their human persecutors. It is a joy to visit the gardens in spring, as much to hear the melody of the birds as to look at the strange and lovely vegetable forms. On a June evening with a pure sunny sky, when the air is elastic after rain, how it rings and palpitates with the fine sounds that people it, and which seem infinite in variety! Has England, burdened with care and long estranged from Nature, so many sweet voices left? What aerial chimes are those wafted from the leafy turret of every tree? What clear, choral songs--so wild, so glad? What strange instruments, not made with hands, so deftly touched and soulfully breathed upon? What faint melodious murmurings that float around us, mysterious and tender as the lisping of leaves? Who could be so dull and exact as to ask the names of such choristers at such a time! Earthly names they have, the names we give them, when they visit us, and when we write about them in our dreary books; but, doubtless, in their brighter home in cloudland they are called by other more suitable appellatives. Kew is exceptionally favoured for the reason

202 BIRDS IN TOWN AND VILLAGE

mentioned, but birds are also abundant where there are no hired men with red waistcoats and brass buttons to watch over their safety. Why do they press so persistently around us; and not in London only, but in every town and village, every house and cottage in this country? Why are they always waiting, congregating as far from us as the depth of garden, lawn, or orchard will allow, yet always near as they dare to come? It is not sentiment, and to be translated into such words as these: "Oh man, why are you unfriendly towards us, or else so indifferent to our existence that you do not note that your children, dependants, and neighbours cruelly persecute us? For we are for peace, and knowing you for the lord of creation, we humbly worship you at a distance, and wish for a share in your affection." No; the small, bright soul which is in a bird is incapable of such a motive, and has only the lesser light of instinct for its guide, and to the birds' instinct we are only one of the wingless mammalians inhabiting the earth, and with the cat and weasel are labelled "dangerous," but the ox and horse and sheep have no such label. Even our larger, dimmer eyes can easily discover the

attraction. Let any one, possessing a garden in the suburbs of London, minutely examine the foliage at a point furthest removed from the house, and he will find the plants clean from insects; and as he moves back he will find them increasingly abundant until he reaches the door. Insect life is gathered thickly about us, for that birdless space which we have made is ever its refuge and safe camping ground. And the birds know. One came before we were up, when cat and dog were also sleeping, and a report is current among them. Like ants when a forager who has found a honey pot returns to the nest, they are all eager to go and see and taste for themselves. Their country is poor, for they have gathered its spoils, and now this virgin territory sorely tempts them. To those who know a bird's spirit it is plain that a mere suspension of hostile action on our part would have the effect of altering their shy habits, and bringing them in crowds about us. Not only in the orchard and grove and garden walks would they be with us, but even in our house. The robin, the little bird "with the red stomacher," would be there for the customary crumbs at meal-time, and many dainty

fringilline pensioners would keep him company. And the wren would be there, searching diligently in the dusty angles of cornices for a savoury morsel; for it knows, this wise little Kitty Wren, that "the spider taketh hold with her hands, and is in king's palaces"; and wandering from room to room it would pour forth many a gushing lyric --a sound of wildness and joy in our still interiors, eternal Nature's message to our hearts.

Who delights not in a bird? Yet how few among us find any pleasure in reading of them in natural history books! The living bird, viewed closely and fearless of our presence, is so much more to the mind than all that is written--so infinitely more engaging in its spontaneous gladness, its brilliant vivacity, and its motions so swift and true and yet so graceful! Even leaving out the melody, what a charm it would add to our homes if birds were permitted to take the part there for which Nature designed them--if they were the "winged wardens" of our gardens and houses as well as of our fields. Bird-biographies are always in our bookcases; and the bird-form meets our sight everywhere in decorative art Eastern and Western; for its aerial beauty is without

---

MOORHENS IN HYDE PARK 205

parallel in nature; but the living birds, with the exception of the unfortunate captives in cages, are not with us.

A robin redbreast in a cage
Puts all heaven in a rage,

sings Blake prophet and poet; and for "robin redbreast" I read every feathered creature endowed with the marvellous faculty of flight. Wild, and loving their safety and liberty, they keep at a distance, at the end of the garden or in the nearest grove, where from their perches they suspiciously watch our movements, always waiting to be encouraged, waiting to feed on the crumbs that fall from our table and are wasted, and on the blighting insects that ring us round with their living multitudes.

THE EAGLE AND THE CANARY

ONE week-day morning, following a crowd of well-dressed people, I presently found myself in a large church or chapel, where I spent an hour very pleasantly, listening to a great man's pulpit eloquence. He preached about genius. The subject was not suggested by the text, nor did it have any close relation with the other parts, of his discourse; it was simply a digression, and, to my mind, a very delightful one. He began about the restrictions to which we are all more or less subject, the aspirations that are never destined to be fulfilled, but are mocked by life's brevity. And it was at this point that--probably thinking of his own case--he branched off into the subject of genius; and proceeded to show that a man possessing that divine quality finds existence a

206

THE EAGLE AND THE CANARY 207

much sadder affair than the ordinary man; the reason being that his aspirations are so much loftier than those of other minds, the difference between his ideal and reality must be correspondingly greater in his case. This was obvious--almost a truism; but the illustration by means of which he brought it home to his hearers was certainly born of poetic imagination. The life of the ordinary person he likened to that of the canary in its cage. And here, dropping his lofty didactic manner, and --if I may coin a word--smalling his deep, sonorous voice, to a thin reedy treble, in imitation of the tenuous fringilline pipe, he went on with lively language, rapid utterance, and suitable brisk movements and gestures, to describe the little lemon-coloured housekeeper in her gilded cage. Oh, he cried, what a bright, busy bustling life is hers, with so many things to occupy her time! how briskly she hops from perch to perch, then to the floor, and back from floor to perch again! how often she drops down to taste the seed in her box, or scatter it about her in a little shower! how curiously, and turning her bright eyes critically this way and that, she listens to every new sound and regards every ob-

ject of sight! She must chirp and sing, and hop from place to place, and eat and drink, and preen her wings, and do at least a dozen different things every minute; and her time is so fully taken up that the narrow limits confining her are almost forgotten--the wires that separate her from the great world of wind-tossed woods, and of blue fields of air, and the free, buoyant life for which her instincts and faculties fit her, and which, alas! can never more be hers.

All this sounded very pretty, as well as true, and there was a pleased smile on every face in the audience.

Then the rapid movements and gestures ceased, and the speaker was silent. A cloud came over his rough-hewn majestic visage; he drew himself up, and swayed his body from side to side, and shook his black gown, and lifted his arms, as their plumed homologues are lifted by some great bird, and let them fall again two or three times; and then said, in deep measured tones, which seemed to express rage and despair, "But did you ever see the eagle in his cage?"

The effect of the contrast was grand. He shook himself again, and lifted and dropped his

arms again, assuming, for the nonce, the peculiar aquiline slouch; and there before us stood the mighty bird of Jove, as we are accustomed to see it in the Zoological Gardens; its deep-set, desolate eyes looking through and beyond us; ruffling its dark plumage, and lifting its heavy wings as if about to scorn the earth, only to drop them again, and to utter one of those long dreary cries which seem to protest so eloquently against a barbarous destiny. Then he proceeded to tell us of the great raptor in its life of hopeless captivity; his stern, rugged countenance, deep bass voice, and grand mouth-filling polysllables suiting his subject well, and making his description seem to our minds a sombre magnificent picture never to be forgotten--at all events, never by an ornithologist.

Doubtless this part of his discourse proved eminently pleasing to the majority of his hearers, who, looking downwards into the depths of their own natures, would be able to discern there a glimmer, or possibly more than a glimmer of that divine quality he had spoken of, and which was, unhappily for them, not recognized by the world at large; so that, for the moment, he was ad-

---

dressing a congregation of captive eagles, all mentally ruffling their plumage and flapping their pinions, and uttering indignant screams of protest against the injustice of their lot.

The illustration pleased me for a different reason, namely, because, being a student of bird-life, his contrasted picture of the two widely different kinds, when deprived of liberty, struck me as being singularly true to nature, and certainly it could not have been more forcibly and picturesquely put. For it is unquestionably the fact that the misery we inflict by tyrannously using the power we possess over God's creatures, is great in proportion to the violence of the changes of condition to which we subject our prisoners; and while canary and eagle are both more or less aerial in their mode of life, and possessed of boundless energy, the divorce from nature is immeasurably greater in one case than in the other. The small bird, in relation to its free natural life, is less confined in its cage than the large one. Its smallness, perching structure, and restless habits, fit it for continual activity, and its flitting, active life within the bars bears some resemblance except in the great matter of flight, to its life in a

---

state of nature. Again, its lively, curious, and extremely impressible character, is in many ways an advantage in captivity; every new sound and sight, and every motion, however slight, in any object or body near it, affording it, so to speak, something to think about. It has the further advantage of a varied and highly musical language; the frequent exercise of the faculty of singing, in birds, with largely developed vocal organs, no doubt reacts on the system, and contributes not a little to keep the prisoner healthy and cheerful.

On the other hand, the eagle, on account of its structure and large size, is a prisoner indeed, and must languish with all its splendid faculties and importunate impulses unexercised. You may gorge it with gobbets of flesh until its stomach cries, "Enough"; but what of all the other organs fed by the stomach, and their correlated faculties? Every bone and muscle and fibre, every feather and scale, is instinct with an energy which you cannot satisfy, and which is like an eternal hunger. Chain it by the feet, or place it in a cage fifty feet wide--in either case it is just as miserable. The illimitable fields of thin cold air, where it outrides the winds and soars exulting be-

---

yond the clouds, alone can give free space for the display of its powers and scope to its boundless energies. Nor to the power of flight alone, but also to a vision formed for sweeping wide horizons, and perceiving objects at distances which to short-sighted man seem almost miraculous. Doubtless, eagles, like men, possess some adaptiveness, else they would perish in their enforced inactivity, swallowing without hunger and assimilating without pleasure the cold coarse flesh we give them. A human being can exist, and even be tolerably cheerful, with limbs paralyzed and hearing gone; and that, to my mind, would be a parallel case to that of the eagle deprived of its liberty and of the power to exercise its flight, vision, and predatory instincts.

As I sit writing these thoughts, with a cage containing four canaries on the table before me, I cannot help congratulating these little prisoners on their comparatively happy fate in having been born, or hatched, finches and not eagles. And yet albeit I am not responsible for the restraint which has been put upon them, and am not their owner, being only a visitor in the house, I am troubled with some uncomfortable feelings concerning their

---

condition--feelings which have an admixture of something like a sense of shame or guilt, as if an injustice had been done, and I had stood by consenting. I did not do it, but we did it. I remember Matthew Arnold's feeling lines on his dead canary, "Poor Matthias," and quote:

Yet, poor bird, thy tiny corse
Moves me, somehow, to remorse;
Something haunts my conscience, brings
Sad, compunctious visitings.
Other favourites, dwelling here,
Open lived with us, and near;
Well we knew when they were glad
Plain we saw if they were sad;
Sympathy could feel and show
Both in weal of theirs and woe.

Birds, companions more unknown,
Live beside us, but alone;
Finding not, do all they can,
Passage from their souls to man.
Kindness we bestow and praise,
Laud their plumage, greet their lays;
Still, beneath their feathered breast

Stirs a history unexpressed.
Wishes there, and feeling strong,
Incommunicably throng;
What they want we cannot guess.

This, as poetry, is good, but it does not precisely fit my case; my "compunctious visitings" being distinctly different in origin and character from the poet's. He--Matthew Arnold--is a poet, and the author of much good verse, which I appreciate and hold dear. But he was not a naturalist--all men cannot be everything. And I, a naturalist, hold that the wishes, thronging the restless little feathered breast are not altogether so incommunicable as the melodious mourner of "Poor Matthias" imagines. The days--ay, and years--which I have spent in the society of my feathered friends have not, I flatter myself, been so wasted that I cannot small my soul, just as the preacher smalled his voice, to bring it within reach of them, and establish some sort of passage.

And so, thinking that a little more knowledge of birds than most people possess, and consideration for them--for I will not be so harsh to speak of justice--and time and attention given to their

wants, might remove this reproach, and silence these vague suggestions of a too fastidious conscience, I have taken the trouble to add something to the seed with which these little prisoners had been supplied. For we give sweetmeats to the child that cries for the moon--an alternative which often acts beneficially--and there is nothing more to be done. Any one of us, even a philosopher, would think it hard to be restricted to dry bread only, yet such a punishment would be small compared with that which we, in our ignorance or want of consideration, inflict on our caged animals--our pets on compulsion. Small, because an almost infinite variety of flavours drawn from the whole vegetable kingdom--a hundred flavours for every one in the dietary which satisfies our heavier mammalian natures--is a condition of the little wild bird's existence and essential to its well-being and perfect happiness. And so, to remedy this defect, I went out into the garden, and with seeding grasses and pungent buds, and leaves of a dozen different kinds, I decorated the cage until it looked less like a prison than a bower. And now for an hour the little creatures have been busy with their varied green

fare, each one tasting half a dozen different leaves every minute, hopping here and there and changing places with his fellows, glancing their bright little eyes this way and that, and all the time uttering gratulatory notes in the canary's conversational tone. And their language is not altogether untranslatable. I listen to one, a pretty pure yellow bird, but slightly tyrannical in his treatment of the others, and he says, or seems to say: "This is good, I like it, only the old leaf is tough; the buds would be better. . . . These are certainly not so good. *I tasted them out of compliment to nature, though they were scarcely palatable. . . .*" No, that was not my own expression; it was said by Thoreau, perhaps the only human a little bird can quote with approval. "This is decidedly bitter--and yet--yes, it does leave a pleasant flavour on the palate. Make room for me there--or I shall make you and let me taste it again. Yes, I fancy I can remember eating something like this in a former state of existence, ages and ages ago." And so on, and so on, until I began to imagine that the whole thing had been put right, and that the uncomfortable feeling would return to trouble me no

more. But at the rate they are devouring their green stuff there will not be a leat, scarcely a stem left in another hour; and then? Why, then they will have the naked wires of their cage all round them to protect them from the cat and for hunger there will be seed in the box.

After all, then, what a little I have been able to do! But I flatter myself that if they were mine I should do more. I never keep captive birds, but if they were given to me, and I could not refuse, I should do a great deal more for them. All my knowledge of their ways and their requirements would teach me how to make their caged existence less unlike the old natural life, than it now is. To begin the ameliorating process, I should place them in a large cage, large enough to allow space for flight, so that they might fly to and fro, a few feet each way, and rest their little feet from continual perching. That would enable them to exercise their most important muscles and experience once more, although in a very limited degree, the old delicious sensation of gliding at will through the void air. The wires of their new cage would be of brass or of some bright metal, and the wooden parts

---

and perches green enamelled, or green variegated with brown and grey, and the roof would be hung with glass lustres, to quiver and sparkle into drops of violet, red, and yellow light, gladdening these little lovers of bright colours; for so we deem them. I should also add gay flowers and berries, crocus and buttercup and dandelion, hips and haws and mountain ash and yellow and scarlet leaves--all seasonable jewellery from woods and hedges and from the orchard and garden. Then would come the heaviest part of my task, which would be to satisfy their continual craving for new tastes in food, their delight in an endless variety. I should go to the great seed-merchants of London and buy samples of all the cultivated seeds of the earth, and not feed them in a trough, or manger, like heavy domestic brutes, but give it to them mixed and scattered in small quantities, to be searched for and gladly found in the sand and gravel and turf on the wide floor of the cage. And, higher up, the wires of their dwelling would be hung with an endless variety of seeded grasses, and sprays of all trees and plants, good, bad, and indifferent. For if the volatile bird dines on no more than twenty dishes every day he

---

loves to taste of a hundred and to have at least a thousand on the table to choose from.

Feeding the birds and keeping the cage always sweet and clean would occupy most, if not the whole of my time. But would that be too much to give if it made me tranquil in my own mind? For it must be noted that I have done all this, mentally and on paper, for my own satisfaction rather than that of the canaries. Birds are not worth much--*to us*. Are not five sparrows sold for three farthings? I have even shot many birds and have felt no compunction. True, they perished before their time, but they did not languish, and being dead there was an end of them; but the caged canaries continuing with us, cannot be dismissed from the mind with the same convenient ease. After all, I begin to think that my imaginary reforms, if carried out, would not quite content me. The "compunctious visitings" would continue still. I look out of the window and see a sparrow on a neighbouring tree, loudly chirruping. And as I listen, trying to find comfort by thinking of the perils which do environ him, his careless unconventional sparrow-music resolves itself into articulate speech, interspersed

---

with occasional bursts of derisive laughter. He knows, this fabulous sparrow, what I have been thinking about and have written. "How would you like it," I hear him saying, "O wise man that knows so much about the ways of birds, if you were shut up in a big cage--in Windsor Castle, let us say--with scores of menials to wait on you and anticipate your every want? That is, I must explain, every want compatible with--ahem!--the captive condition. Would you be happy in your confinement, practising with the dumb-bells, riding up and down the floors on a bicycle and gazing at pictures and filigree caskets and big malachite vases and eating dinners of many, many courses? Or would you begin to wish that you might be allowed to live on sixpence a day--*and earn it*; and even envy the ragged tramp who dines on a handful of half-rotten apples and sleeps in a hay-stack, but is free to come and go, and range the world at will? You have been playing at nature; but Nature mocks you, for your captives thank you not. They would rather go to her without an intermediary, and take a scantier measure of food from her hand, but flavoured as she only can flavour it. Widen your cage,

---

THE EAGLE AND THE CANARY 221

naturalist; replace the little twinkling lustres with sun and moon and milky way; plant forests on the floor, and let there be hills and valleys, rivers and wide spaces; and let the blue pillars of heaven be the wires of your cage, with free entrance to wind and rain; then your little captives will be happy, even happy as I am, in spite of all the perils which do environ me--guns and cats and snares, with wet and fog and hard frosts to come."

And, seeing my error, I should open the cage and let them fly away. Even to death, I should let them fly, for there would be a taste of liberty first, and life without that sweet savour, whether of aerial bird or earth-

bound man, is not worth living.

CHANTICLEER

DURING the month of September I spent several days at a house standing on high ground in one of the pleasantest suburbs of London, commanding a fine view at the back of the breezy, wooded, and not very far-off Surrey hills; and all round, from every window, front and back, such a mass of greenery met the eye, almost concealing the neighbouring houses, that I could easily imagine myself far out in the country. In the garden the omnipresent sparrow, and that always pleasant companion the starling, associated with the thrush, blackbird, green linnet, chaffinch, redstart, wren, and two species of tits; and, better than all these, not fewer than half a dozen robins warbled their autumn notes from early morning until late in the evening. Domestic

222

CHANTICLEER 223

bird-life was also represented by fifteen fowls, and the wise laxity existing in the establishment made these also free of the grounds; for of eyesores and painful skeletons in London cupboards, one of the worst, to my mind, is that unwholesome coop at the back where a dozen unhappy birds are usually to be found immured for life. These, more fortunate, had ample room to run about in, and countless broad shady leaves from which to pick the green caterpillar, and red tortoise-shaped ladybird, and parti-coloured fly, and soft warm soil in which to bathe in their own gallinaceous fashion, and to lie with outstretched wings luxuriating by the hour in the genial sunshine. And having seen their free wholesome life, I did not regard the new-laid egg on the breakfast-table with a feeling of repugnance, but ate it with a relish.

I have said that the fowls numbered fifteen; five were old birds, and ten were chickens, closely alike in size, colour and general appearance. They were not the true offspring of the hen that reared them, but hatched from eggs bought from a local poultry-breeder. As they advanced in age to their teens, or the period in chicken-life corre-

---

224 BIRDS IN TOWN AND VILLAGE

sponding to that in which, in the human species, boy and girl begin to diverge, their tails grew long, and they developed very fine red combs; but the lady of the house, who had been promised good layers when she bought the eggs clung tenaciously to the belief that long arching tails and stately crests were ornaments common to both sexes in this particular breed. By and by they commenced to crow, first one, then two, then all, and stood confessed cockerels. Incidents like this, which are of frequent occurrence, serve to keep alive the exceedingly ancient notion that the sex of the future chick can be foretold from the shape of the egg. As I had no personal interest in the question of the future egg-supply of the establishment, I was not sorry to see the chickens develop into cocks; what did interest me were their first attempts at crowing--those grating sounds which the young bird does not seem to emit, but to wrench out with painful effort, as a plant is wrenched out of the soil, and not without bringing away portions of the lungs clinging to its roots. The bird appears to know what is coming, like an amateur dentist about to extract one of his own double-pronged teeth, and setting his feet

firmly on the ground, and throwing himself well back before an imaginary looking-glass, and with arched-neck, wide-open beak, and rolling eyes, courageously performs the horrible operation. One cannot help thinking that a cockerel brought up without any companions of his own sex and age would not often crow, but in this instance there were no fewer than ten of them to encourage each other in the laborious process of tuning their harsh throats. Heard subsequently in the quiet of the early morning, these first tuning efforts suggested some reflections to my mind, which may not prove entirely without interest to fanciers who aim at something beyond a mere increase in our food-supply in their selecting and refining processes.

To continue my narration. I woke in the morning at my usual time, between three and four o'clock, which is not my getting-up time, for, as a rule, after half an hour or so I sleep again. The waking is not voluntary as far as I know; for although it may seem a contradiction in terms to speak of coming at will out of a state of unconsciousness, we do, in cases innumerable, wake voluntarily, or at the desired time, not perhaps

---

being altogether unconscious when sleeping. If, however, this early waking were voluntary, I should probably say that it was for the pleasure of listening to the crowing of the cocks at that silent hour when the night, so near its end, is darkest, and the mysterious tide of life, prescient of coming dawn, has already turned, and is sending the red current more and more swiftly through the sleeper's veins. I have spent many a night in the desert, and when waking on the wide silent grassy plain, the first whiteness in the eastern sky, and the fluting call of the tinamou, and the perfume of the wild evening primrose, have seemed to me like a resurrection in which I had a part; and something of this feeling is always associated in my mind with the first far-heard notes of Chanticleer.

It was very dark and quiet when I woke; my window was open, with only a lace curtain before it to separate me from the open air. Presently the profound silence was broken. From a distance of fifty or sixty yards away on the left hand came the crow of a cock, soon answered by another further away on the same side, and then, further away still, by a third. Other voices took

---

up the challenge on the right, some near, some far, until it seemed that there was scarcely a house in the neighbourhood at which Chanticleer was not a dweller. There was no other sound. Not for another hour would the sparrows burst out in a chorus of chirruping notes, lengthened or shortened at will, variously inflected, and with a ringing musical sound in some of them, which makes one wonder why this bird, so high in the scale of nature, has never acquired a set song for itself. For there is music in him, and when confined with a singing finch he will sometimes learn its song. Then the robins, then the tits, then the starlings, gurgling, jarring, clicking, whistling, chattering. Then the pigeons cooing soothingly on the roof and window-ledges, taking flight from time to time with sudden, sharp flap, flap, followed by a long, silken sound made by the wings in gliding. At four the cocks had it all to themselves; and, without counting the cockerels (not yet out of school), I could distinctly hear a dozen birds; that is to say, they were near enough for me to listen to their music critically. The variety of sounds they emitted was very great, and, if cocks were selected for their vocal qualities,

---

would have shown an astonishing difference in the musical tastes of their owners. A dozen dogs of as many different breeds, ranging from the boar-hound to the toy terrier, would not have shown greater dissimilarity in their forms than did these cocks in their voices. For the fowl, like the dog, has become an extremely variable creature in the domestic state, in voice no less than in size, form, colour, and other particulars. At one end of the scale there was the raucous bronchial strain produced by the unwieldy Cochin. What a bird is that! Nature, in obedience to man's behests, and smiling with secret satire over her work, has made it ponderous and ungraceful as any clumsy mammalian, wombat, ardvaark, manatee, or hippopotamus. The burnished red hackles, worn like a light mantle over the black doublet of the breast, the metallic dark green sickle-plumes arching over the tail, all the beautiful lines and rich colouring, have been absorbed into flesh and fat for gross feeders; and with these have gone its liveliness and vigour, its clarion voice and hostile spirit and brilliant courage; it is Gallus bankiva degenerate, with dulled brains

---

CHANTICLEER 229

and blunted spurs, and its hoarse crow is a barbarous chant.

And far away at the other end, startling in its suddenness and impetuosity, was a trisyllabic crow, so brief, piercing, and emphatic, that it could only have proceeded from that peppery uppish little bird, the bantam. And of the three syllables, the last, which should be the longest, was the shortest, "short and sharp like the shrill swallow's cry," or perhaps even more like the shrieky bark of an enraged little cur; not a *reveille* and silvern morning song in one, as a crow should be, but a challenge and a defiance, wounding the sense like a spur, and suggesting the bustle and fury of the cockpit.

If this style of crowing was known to Milton, it is perhaps accountable for the one bad couplet in the "Allegro":

While the cock with lively din Scatters the rear of darkness thin.

Someone has said that every line in that incomparable poem brings at least one distinct picture vividly before the mind's eye. The picture

---

230 BIRDS IN TOWN AND VILLAGE

the first line of the couplet I have quoted suggests to ray mind is not of crowing Chanticleer at all, but of a stalwart, bare-armed, blowsy-faced woman, vigorously beating on a tin pan with a stick; but for what purpose--whether to call down a passing swarm of bees, or to summon the chickens to be fed--I never know. It is only my mental picture of a "lively din." As to the second line, all attempts to see the thing described only bring before me clouds and shadows, confusedly rushing about in an impossible way; a chaos utterly unlike the serenity and imperceptible growth of morning, and not a picture at all.

By and by I found myself paying special attention to one cock, about a hundred yards away, or a little more perhaps, for by contrast all the other songs within hearing seemed strangely inferior. Its voice was singularly clear and pure, the last note greatly prolonged and with a slightly falling inflection, yet not collapsing at the finish as such long notes frequently do, ending with a little internal sound or croak, as if the singer had exhausted his breath; but it was perfect in its way, a finished performance, artistic, and, by comparison, brilliant. After once hearing this bird I paid little attention to the others, but after each resounding call I counted the seconds until its repetition. It was this bird's note, on this morning, and not the others, which seemed to bring round me that atmosphere of dreams and fancies I exist in at early cockcrow--dreams and memories, sweet or sorrowful, of old scenes and faces, and many eloquent passages in verse and prose, written by men in other and better days, who lived more with nature than we do now. Such a note as this was, perhaps, in Thoreau's mind when he regretted that there were no cocks to cheer him in the solitude of Walden. "I thought," he says, "that it might be worth while keeping a cockerel for his music merely, as a singing bird. The note of this once wild Indian pheasant is certainly the most remarkable of any bird's, and if they could be naturalized without being domesticated it would soon become the most famous sound in our woods. . . . To walk in a winter morning in a wood where these birds abounded, their native woods, and hear the wild cockerels crow on the trees, clear and shrill for miles over the surrounding coun-

try--think of it! It would put nations on the alert. Who would not be early to rise, and rise earlier and earlier on each successive morning of his life, till he became unspeakably healthy, wealthy, and wise?"

Soon I fell into thinking of one in some ways greater than Thoreau, so unlike the skyey-minded New England prophet and solitary, so much more genial and tolerant, more mundane and lovable; and yet like Thoreau in his nearness to nature. Not only a lover of generous wines--"That mark upon his lip is wine"--and books "clothed in black and red," all natural sights and sounds also "filled his herte with pleasure and solass," and the early crowing of the cock was a part of the minstrelsy he loved. Perhaps when lying awake during the dark quiet hours, and listening to just such a note as this, he conceived and composed that wonderful tale of the "Nun's Priest," in which the whole character of Chanticleer, his glory and his foibles, together with the homely virtues of Dame Partlett, are so admirably set forth.

And longer ago it was perhaps such a note as this, heard in imagination by the cock-loving

---

Athenians, which all at once made them feel so unutterably weary of endless fighting with the Lacedaemonians, and inspired their hearts with such a passionate desire for the long untasted sweets of security and repose. Is it one of my morning fancies merely--for fact and fancy mingle strangely at this still, mysterious hour, and are scarcely distinguishable--or is it related in history that this strange thing happened when all the people of the violet-crowned city were gathered to witness a solemn tragedy, in which certain verses were spoken that had a strange meaning to their war-weary souls? "Those who sleep in the morning in the arms of peace do not start from them at the sound of the trumpet, and nothing interrupts their slumbers but the peaceful crowing of the cock." And at these words the whole concourse was electrified, and rose up like one man, and from thousands of lips went forth a great cry of "Peace! Peace! Let us make peace with Sparta!"

Hark! once more that long clarion call: it is the last time--the very last; for all the others have sung a dozen times apiece and have gone to sleep again. So would this one have done,

---

but cocks, like minstrels among men, are vain creatures, and some kind officious fairy whispered in his ear that there was an appreciative listener hard by, and so to please me he sang, just one stave more.

Lying and listening in the dark, it seemed to me that there were two opposite qualities commingled in the sound, with an effect analogous to that of shadow mingling with and chastening light at eventide. First, it was strong and clear, full of assurance and freedom, qualities admirably suited to the song of a bird of Chanticleer's disposition; a lusty, ringing strain, not sung in the clouds or from a lofty perch midway between earth and heaven, but with feet firmly planted on the soil, and earthly; and compared with the notes of the grove like a versified utterance of Walt Whitman compared with the poems of the true inspired children of song--Blake, Shelley, Poe. Earthly, but not hostile and eager; on the contrary, leisurely, *peaceful* even dreamy, with a touch of tenderness which brings it into relationship with the more aerial tones of the true singers; and this is the second quality I spoke of, which gave a charm to this note and made it

---

CHANTICLEER 235

seem better than the others. This is partly the effect of distance, which clarifies and softens sound, just as distance gives indistinctness of outline and ethereal blueness to things that meet the sight. To objects beautiful in themselves, in graceful lines and harmonious proportions and colouring, the haziness imparts an additional grace; but it does not make beautiful the objects which are ugly in themselves, as, for instance, an ugly square house. So in the etherealizing effect of distance on sound, when so loud a sound as the crowing of a strong-lunged cock becomes dreamy and tender at a distance of one hundred yards, there must be good musical elements in it to begin with. I do not remark this dreaminess in the notes of other birds, some crowing at an equal distance, others still further away. All natural music is heard best at a distance; like the chiming of bells, and the music of the flute, and the wild confused strains of the bagpipes, for among artificial sounds these come the nearest to those made by nature. The "shrill sharps" of the thrush must be softened by distance to charm; and the skylark, when close at hand, has both shrill and harsh sounds scarcely pleasing. He must mount high before you can appreciate his merit. I do not recommend any one to keep a caged cock in his study for the sake of its music, crow it never so well.

To return to the ten cockerels; they did not crow very much, and at first I paid little attention to them. After a few days I remarked that one individual among them was rapidly acquiring the clear vigorous strain of the adult bird. Compared with that fine note which I have described, it was still weak and shaky, but in shape it was similar, and the change had come while its brethren were still uttering brief and harsh screeches as at the beginning. Probably, where there is a great mixture of varieties, it is the same with the fowl as with man in the diversity of the young, different ancestral characters appearing in different members of the same family. This cockerel was apparently the musical member, and promised in a short time to rival his neighbour. Having heard that it was intended to keep one of the cockerels to be the parent of future broods, I began to wonder whether the prize in the lottery--to wit, life and a modest harem--would fall to this fine singer or not. The odds were that

his musical career would be cut short by an early death, since the ten birds were very much alike in other respects, and I felt perfectly sure that his superior note would weigh nothing in the balance. For when has the character of the voice influenced a fancier in selecting? Never I believe, odd as it seems. I have read a very big book on the various breeds of the fowl, but the crowing of the cock was not mentioned in it. This would not seem so strange if fanciers had invariably looked solely to utility, and their highest ambition had ended at size, weight and quality of flesh, early maturity, hardihood, and the greatest number of eggs. This has not been the case. They possess, like others, the love of the beautiful, artificial as their standards sometimes appear; and there are breeds in which beauty seems to have been the principal object, as, for instance, in several of the gold and silver spangled and pencilled varieties. But, besides beauty of plumage, there are other things in the fowl worthy of being improved by selection. One of these has been cultivated by man for thousands of years, namely, the combative spirit and splendid courage of the male bird. But there is a spirit

---

238 BIRDS IN TOWN AND VILLAGE

abroad now which condemns cock-fighting, and to continue selecting and breeding cocks solely for their game-points seems a mere futility. The energy and enthusiasm expended in this direction would be much better employed in improving the bird's vocal powers.

The morning song of the cock is a sound unique in nature, and of all natural sounds it is the most universal. "All climates agree with brave Chanticleer. He is more indigenous even than the natives. His health is ever good; his lungs are sound; his spirits never flag." He is a pet bird among tribes that have never seen the peacock, goose, and turkey. In tropical countries where the dog becomes dumb, or degenerates into a mere growler, his trumpet never rusts. It is true that he was cradled in the torrid zone, yet in all Western lands, where he "shakes off the powdery snow," with vigorous wings, his voice sounds as loud and inspiriting as in the hot jungle. Pale-faced Londoners, and blacks, and bronzed or painted barbarians, all men all the world over, wake at morn to the "peaceful crowing of the cock," just as the Athenians woke of old, and the nations older still. It is not, therefore,

---

CHANTICLEER 239

strange that this song has more associations for man than any other sound in nature. But, apart from any adventitious claims to our attention, the sound possesses intrinsic merits and pleases for its own sake. In our other domestic birds we have, with regard to this point, been unfortunate. We have the gobbling of turkeys, and the hoarse, monotonous come back of the guinea-fowl, screaming of peacocks and geese, and quacking, hissing, and rasping of mallard and mus-covy. Above all these sounds the ringing, lusty, triumphant call of Chanticleer, as the far-reaching toll of the bell-bird sounds above the screaming and chattering of parrots and toucans in the Brazilian forest. A fine sound, which in spite of many changes of climate and long centuries of domestication still preserves that forest-born character of wildness, which gives so great a charm to the language of many woodland gallinaceous birds. As we have seen, it is variable, and in some artificial varieties has been suffered to degenerate into sounds harsh and disagreeable; yet it is plain that an improved voice in a beautiful breed would double the bird's value from an aesthetic point of view. As things now are, the

---

fine voices are in a very small minority. Some bad voices in artificial breeds, i.e., those which, like the Brahma and Cochin, diverge most widely from the original type--are perhaps incurable, like the carrion crow's voice; for that bird will probably always caw harshly in spite of the musical throat which anatomists find in it. We can only listen to our birds, and begin experimenting with those already possessed of shapely notes and voices of good quality.

I am not going to be so ill-mannered as to conclude without an apology to those among us who under no circumstances can tolerate the crowing of the cock. It is true that I have not been altogether unmindful of their prepossessions, and have freely acknowledged in divers places that Chanticleer does not always please, and that there is abundant room for improvement; but if they go further than that, if for them there exists not on this round globe a cock whose voice would fail to irritate, then I have not shown consideration enough, and something is still owing to their feelings, which are very acute. It is possible that one of these sensitive persons may take up my book, and, attracted by its title, dip into this paper, hoping to find in it a practical suggestion for the effectual muzzling of the obnoxious bird. The only improvement which would fall in with such a one's ideas on the subject of cockcrowing would be to improve this kind of natural music out of existence. Naturally the paper would disappoint him; he would be grieved at the writer's erroneous views. I hope that his feelings would take no acuter form. I have listened to a person, usually mild-mannered, denouncing a neighbour in the most unmeasured terms for the crime of keeping a crowing cock. If the cock had been a non-crower, a silent member, it would have been different: he would hardly have known that he had a neighbour. There is a very serious, even a sad, side to this question. Mr. Sully maintains that as civilization progresses, and as we grow more intellectual, all noise, which is pleasing to children and savages, and only exhilarates their coarse and juvenile brains, becomes increasingly intolerable to us. What unfortunate creatures we then are! We have got our pretty rattle and are now afraid that the noise it makes is going to be the death of us. But what is noise? Will any two highly intellectual beings agree as

to the particular sound which produces the effect of rusty nails thrust in among the convolutions of the brain? Physicians are continually discovering new forms of nervous maladies, caused by the perpetual hurry and worry and excitement of our modern life; and perhaps there is one form in which natural sounds, which being natural should be agreeable, or at any rate innocent, become more and more abhorrent. This is a question which concerns the medical journals; also, to some extent, those who labour to forecast the future. Happily, all our maladies are thrown off, sooner or later, if they do not kill us; and we can cheerfully look forward to a time when the delicate chords in us shall no longer be made to vibrate "like sweet bells jangled out of tune and harsh" to any sound in nature, and when the peaceful crowing of the cock shall cease to madden the early waker. For, whatever may be the fate awaiting our city civilization, brave Chanticleer, improved as to his voice or not,

will undoubtedly still be with us.

IN AN OLD GARDEN

A SUNNY morning in June--a golden day among days that have mostly a neutral tint; a large garden, with no visible houses beyond, but green fields and unkept hedges and great silent trees, oak and ash and elm--could I wish, just now, for a more congenial resting-place, or even imagine one that comes nearer to my conception of an earthly paradise? It is true that once I could not drink deeply enough from the sweet and bitter cup of wild nature, and loved nature best, and sought it gladly where it was most savage and solitary. But that was long ago. Now, after years of London life, during which I have laboured like many another "to get a wan pale face," with perhaps a wan pale mind to match, that past wildness would prove too potent and sharp a tonic; unadulterated nature would startle and oppress me with its rude desolate aspect, no longer familiar. This softness of a well-cultivated earth, and unbroken verdure of foliage in many shades, and harmonious grouping and blending of floral hues, best suit my present enervated condition. I had, I imagine, a swarter skin and firmer flesh when I could ride all day over great summer-parched plains, where there was not a bush that would have afforded shelter to a mannikin, and think that I was having a pleasant journey. The cloudless sky and vertical sun--how intolerable they would now seem, and scorch my brain and fill my shut eyes with dancing flames! At present even this mild June sun is strong enough to make the old mulberry tree on the lawn appear grateful. It is an ancient, rough-barked tree, with wide branches, that droop downwards all round, and rest their terminal leaves on the sward; underneath it is a natural tent, or pavilion, with plenty of space to move about and sling a hammock in. Here, then, I have elected to spend the hottest hours of my one golden day, reading, dreaming, listening at intervals to the fine bird-sounds that have a.

IN AN OLD GARDEN 345

medicinal and restorative effect on the jarred and wounded sense.

From the elms hard by comes a subdued, airy prattle of a few sparrows. It is rather pleasant, something like a low accompaniment to the notes of the more tuneful birds; the murmurous music of a many-stringed instrument, forming the indistinct ground over which runs the bright embroidery of clear melodious singing.

This morning, while lying awake from four to five o'clock, I almost hated the sparrows, they were there in such multitudes, and so loud and persistent sounded their jangling through the open window. It set me thinking of the England of the future--of a time a hundred years hence, let us say--when there will remain with us only two representatives of feral life--the sparrow and the house-fly. Doubtless it will come, unless something happens; but, doubtless, it will not continue. It will still be necessary for a man to kill something in order to be happy; and the sportsmen of that time, like great Gambetta, in the past, will sit in the balconies, popping with pea-rifles at the sparrows until not one is left to twitter. Then will come the turn of the untamed

---

and untamable fly; and he will afford good sport if hunted a la Domitain, with fine, needle-tipped paper javelins, thrown to impale him on the wall.

One of our savants has lately prophesied that the time will come when only the microscopic organisms will exist to satisfy the hunting instinct in man. How these small creatures will be taken he does not tell us. Perhaps the hunters will station themselves round a table with a drop of preserved water on its centre, made large and luminous by means of a ray of magnifying light. When that time comes the amoeba--that "wandering Jew," as an irreverent Quarterly Reviewer has called it--will lose its immortality, and the spry rotifer will fall a victim to the infinitesimal fine bright arrows of the chase. A strange quarry for men whose paeliolithic progenitors hunted the woolly mastodon and many-horned rhinoceros and sabre-toothed tiger!

That sad day of very small things for the sportsman is, however, not near, nor within measurable distance; or, so it seemed to me when, an hour ago, I strolled round the garden, curiously peering into every shrub, to find the visible and comparatively noble insect-life in great

---

abundance. Beetles were there--hard, round, polished, and of various colours, like sea-worn pebbles on the beach; and some, called ladybirds in the vernacular, were bound like the books that Chaucer loved in black and red. And the small gilded fly, not less an insect light-headed, a votary of vain delights, than in the prehistoric days when a white-headed old king, discrowned and crazed, railed against sweet Nature's liberty. And ever waiting to welcome this inconstant lover (with falces) there sits the solitary geometric spider, an image and embodiment of patience, not on a monument, but a suspended wheel of which he is himself the hub; and so delicately fashioned are the silver spokes thereof, radiating from his round and gem-like body, and the rings, concentric tire within tire, that its exceeding fineness, like swift revolving motion, renders it almost invisible. Caterpillars, too, in great plenty--miniature porcupines with fretful quills on end, and some naked even as they came into the world. This one, called the earth-measurer, has drunk himself green with chlorophyll so as to escape detection. Vain precaution! since eccentric motion betrays him to keen avian eyes, when, like

---

the traveller's snake, he erects himself on the tip of his tail and sways about in empty space, vaguely feeling for something, he knows not what. And the mechanical tortrix that rolls up a leaf for garment and food, and preys on his own case and shelter until he has literally eaten himself stark naked; after which he rolls up a second leaf, and so on progressively. Thus in his larval life does he symbolize some restless nation that makes itself many successive constitutions and forms of government, in none of which it abides long; but afterwards some higher thing, when he rests motionless, in form like a sarcophagus, whence the infolded life emerges to haunt the twilight--a grey ghost moth. There is no end to rolled-up leaves, and to the variety of creatures that are housed in them; for, just as the "insect tribes of human kind" in all places and in all ages, while seeking to improve their condition, independently hit on the same means and inventions, so it is with these small six-legged people; and many species in many places have found out the comfort and security of the green cylinder.

So many did I open that I at last grew tired

---

of the process, like a man to whom the post has brought too many letters; but there was one--the last I opened--the living active contents of which served to remind me that some insects are unable to make a cylinder for themselves, having neither gum nor web to fasten it with, and yet they will always find one made by others to shelter themselves in. Here were no fewer than six unbeautiful creatures, brothers and sisters, hatched from eggs on which their parent earwig sat incubating just like an eagle or dove or swallow, or, better still, like a pelican; for in the end did she not give of her own life-fluid to nourish her children? Unbeautiful, yet not without a glory superior to that of the Purple Emperor, and the angelic blue Morpho, and the broad-winged Ornithoptera, that caused an illustrious traveller to swoon with joy at the sight of its supreme loveliness. Du Maurier has a drawing of a little girl in a garden gazing at two earwigs racing along a stem. "I suppose," she remarks interrogatively to her mamma, "that these are Mr. and Mrs. Earwig?" and on being answered affirmatively, exclaims, "What could they have seen in each other?" What they saw was blue

---

blood, or something in insectology corresponding to it. The earwig's lustre is that of antiquity. He existed on earth before colour came in; and colour is old, although not so old as Nature's unconscious aestheticism which, in the organic world, is first expressed in beauty of form. It is long since the great May flies, large as swifts, had their aerial cloudy dances over the vast everglades and ancient forests of ferns; and when, on some dark night, a brilliant Will-o'-the-wisp rose and floated above the feathery foliage, drawn in myriads to its light, they revolved about it in an immense mystical wheel, misty-white, glistening, and touched with prismatic colour. Floating fire and wheel were visible only to the stars, and the wakeful eyes of giant scaly monsters lying quiescent in the black waters below; but they were very beautiful nevertheless. The modest earwig was old on the earth even then; he dates back to the time, immeasurably remote, when scorpions possessed the earth, and taught him to frighten his enemies with a stingless tail--that curious antique little tail which has not yet forgot its cunning.

Greater than all these inhabitants of the garden, ancient or modern by reason of their

---

numbers, which is the sign of predominance, are the small wingless people that have colonies on every green stem and under every green leaf.

These are the true generators of that heavenly sweat, or saliva of the stars, concerning which Pliny the Younger wrote so learnedly. And they are many tribes--green, purple, brown, isabel-line; but all are one nation, and sacred to that fair god whom the Carian water-nymph loved not wisely but too well. For, albeit the children of an ancient union, they marry not, nor are given in marriage, yet withal multiply exceedingly, so that one (not two) may in a single season produce a billion. And at last when autumn comes, won back from the cold god to his hot mother, they know love and wedlock, and die like all married things. These are the Aphides--sometimes unprettily called plant-lice, and vaguely spoken of by the uninformed as "blight"--and they nourish themselves on vegetable juices, that thin green blood which is the plant's life.

This, then, is the fruit which the birds have, come to gather. In June is their richest harvest; it is more bountiful than September, when apples redden, and grapes in distant southern lands are

---

gathered for the wine-press. In yon grey wall at the end of the lawn, just above the climbing rose-bush, there are now seven hungry infants in one small cradle, each one, some one says, able to consume its own weight of insect food every day. I am inclined to believe that it must be so, while trying to count the visits paid to the nest in one hour by the parent tits--those small tits that do the gardener so much harm! We know, on good authority, that the spider has a "nutty flavour"; and most insects in the larval stage afford succulent and toothsome, or at all events beaksome, morsels. These are, just now, the crimson cherries, purple and yellow plums, currants, red, white, and black--and sun-painted peaches, asking in their luscious ripeness for a mouth to melt in, that fascinate finch and flycatcher alike, and make the starlings smack their horny lips with a sound like a loving kiss.

Not that I care, or esteem birds for what they eat or do not eat. With all these creatures that are at strife among themselves, and that birds prey upon, I am at peace, even to the smallest that are visible--the red spider which is no spider; and the minute gossamer spider clinging to

---

the fine silvery hairs of the flying summer; and the coccus that fall from the fruit trees to float on their buoyant cottony down--a summer snow. Fils de la Vierge are these, and sacred. The man who can needlessly set his foot on a worm is as strange to my soul as De Quincey's imaginary Malay, or even his "damned crocodile." The worm that one sees lying bruised and incapable on the gravel walk has fallen among thieves. These little lives do me good and not harm. I smell the acid ants to strengthen my memory. I know that if I set an overturned cockchafer on his legs three sins shall be forgiven me; that if I am kindly tolerant of the spider that drops accidentally on my hand or face, my purse shall be mysteriously replenished. At the same time, one has to remember that such sentiments, as a rule, are not understood by those who have charge over groves and gardens, whose minds are ignorant and earthy, or, as they would say, practical. Of the balance of nature they know and care naught, nor can they regard life as sacred; it is enough to know that it is or may be injurious to their interests for them to sweep it away. The small thing that has been flying about and uttering

---

musical sounds since April may, when July comes, devour a certain number of cherries. Nor is even this plea needed. If it is innocent for the lower creatures to prey upon one another, it cannot be less innocent for man to destroy them indiscriminately, if it gives him any pleasure to do so. It is idle to go into such subtle questions with those who have the power to destroy; if their hands are to be restrained it is not by appealing to feelings which they do not possess, but to their lower natures--to their greed and their cunning. For the rest of us, for all who have conquered or outgrown the killing instinct, the impartiality that pets nothing and persecutes nothing is doubtless man's proper attitude towards the inferior animals; a godlike benevolent neutrality; a keen and kindly interest in every form of life, with indifference as to its ultimate destiny; the softness which does no wrong with the hardness that sees no wrong done.

To return to the birds. The starlings have kissed like lovers, and fluttered up vertically on their short wings, trying to stream like eagles, only to return to the trees once more and sit there chattering pleasant nothings; at intervals

---

throwing out those soft, round, modulated whistled notes, just as an idle cigarette-smoker blows rings of blue smoke from his lips; and now they have flown away to the fields so that I can listen to the others.

A thrush is making music on a tall tree beyond the garden hedge, and I am more grateful for the distance that divides us than for the song; for, just now, he does not sing so well as sometimes of an evening, when he is most fluent, and a listener, deceived by his sweetness and melody, writes to the papers to say that he has heard the nightingale. Just now his song is scrappy, composed of phrases that follow no order and do not fit or harmonize, and is like a poor imitation of an inferior mocking-bird's song.

Between the scraps of loud thrush-music I listen to catch the thin, somewhat reedy sound of a yellow-hammer singing in the middle of the adjoining grassy field. It comes well from the open expanse of purpling grass, and reminds me of a favourite grasshopper in a distant sunny land. O happy grasshopper! singing all day in the trees and tall herbage, in a country where every village urchin is not sent afield to "study natural his-

tory" with green net and a good store of pins, shall I ever again hear thy breezy music, and see thee among the green leaves, beautiful with steel-blue and creamy-white body, and dim purple over and vivid red underwings?

The bird of the pasture-land is singing still, perhaps, but all at once I have ceased to hear him, for something has come to lift me above his low grassy level, something faint and at first only the suspicion of a sound; then a silvery lisping, far off and aerial, touching the sense as lightly as the wind-borne down of dandelion.

If any place for any soul there be Disrobed and disentrammelled,

doubtless it is from such a place and such a soul that this sublimated music falls. The singer, one can imagine, has never known or has forgotten earth; and if it is visible to him, how small it must seem from that altitude, "spinning like a fretful midge" beneath him in the vast void I

It is the lark singing in the blue infinite heaven, at this distance with something ethereal and heavenly in his voice; but now the wide circling

wings that brought him for a few moments within hearing, have borne him beyond it again; and missing it, the sunshine looks less brilliant than before, and all other bird-voices seem by comparison dull and of the earth.

Certainly there is nothing spiritual in the song of the chaffinch. There he sits within sight, motionless, a little bird-shaped automaton, made to go off at intervals of twelve or thirteen seconds; but unfortunately one hears with the song the whirr and buzz of the internal machinery. It is not now as in April, when it is sufficient in a song that it shall be joyous; in the leafy month, when roses are in bloom, one grows critical, and asks for sweetness and expression, and a better art than this vigorous garden singer displays in that little double flourish with which he concludes his little hurry-scurry lyric. He has practised that same flourish for five thousand years--to be quite within the mark--and it is still far from perfect, still little better than a kind of musical sneeze. So long is art!

Perhaps in some subtle way, beyond the psychologist's power to trace, he has become aware of my opinion of his performance--the unspoken

---

detraction which yet affects its object; and, feeling hurt in his fringilline *amour propre,* he has all at once taken himself off. Never mind; a better singer has succeeded him. I have heard and seen the little wren a dozen times to-day; now he has come to the upper part of the tree I am lying under, and although so near his voice sounds scarcely louder than before. This is also a lyric, but of another kind. It is not plaintive, nor passionate; nor is it so spontaneous as the warbling of the robin--that most perfect feathered impressionist; nor is it endeared to me by early associations since I listened in boyhood to the songs of other wrens. In what, then, does its charm consist? I do not know. Certainly it is delicate, and may even be described as brilliant, in its limited way perfect, and to other greater songs like the small pimpernel to a poppy or a hollyhock. Unambitious, yet finished, it has the charm of distinction. The wren is the least self-conscious of our singers. Somewhere among the higher green translucent leaves the little brown barred thing is quietly sitting, busy for the nonce about nothing, dreaming his summer dream, and unknowingly telling it aloud. When shall

---

we have symbols to express as perfectly our summer-feeling--our dream ?

That small song has served to remind me of two small books I brought into the garden to read--the works of two modern minor poets whose "wren-like warblings," I imagined, would suit my mood and the genial morning better than the stirring or subtle thoughts of greater singers. Possibly in that I was mistaken; for there until now lie the books neglected on a lawn chair within reach of my hand. The chair was dragged hither half-an-hour ago by a maiden all in white, who appeared half inclined to share the mulberry shade with me. She did not continue long in that mind. In a lively manner, she began speaking of some trivial thing; but after a very few moments all interest in the subject evaporated, and she sat humming some idle air, tapping the turf with her fantastic shoe. Presently she picked up one of my books, opened it at random and read a line or two, her vermilion under-lip curling slightly; then threw it down again, and glanced at me out of the corners of her eyes; then hummed again, and finally became silent, and sat bending forward a little, her dark lustrous eyes gazing with

strange intentness through the slight screen of foliage into the vacant space beyond. What to see? The poet has omitted to tell us to what the maiden's fancy lightly turns in spring. Doubtless it turns to thoughts of something real. Life is real; so is passion--the quickening of the blood, the wild pulsation. But the pleasures and pains of the printed book are not real, and are to reality like Japanese flowers made of coloured bits of tissue paper to the living fragrant flowers that bloom to-day and perish to-morrow; they are a simulacrum, a mockery, and present to us a pale phantasmagoric world, peopled with bloodless men and women that chatter meaningless things and laugh without joy. The feeling of unreality affects us all at times, but in very different degrees. And perhaps I was too long a doer, herding too much with narrow foreheads, drinking too deeply of the sweet and bitter cup, to experience that pure unfailing delight in literature which some have. Its charm, I fancy, is greatest to those in whom the natural man, deprived in early life of his proper aliment, grows sickly and pale, and perishes at last of inanition. There is ample room then for the latter higher growth--

---

the unnatural cultivated man. Lovers of literature are accustomed to say that they find certain works "helpful" to them; and doubtless, being all intellect, they are right. But we, the less highly developed, are compounded of two natures, and while this spiritual pabulum sustains one, the other and larger nature is starved; for the larger nature is earthly, and draws its sustenance from the earth. I must look at a leaf, or smell the sod, or touch a rough pebble, or hear some natural sound, if only the chirp of a cricket, or feel the sun or wind or rain on my face. The book itself may spoil the pleasure it was designed to give me, and instead of satisfying my hunger, increase it until the craving and sensation of emptiness becomes intolerable. Not any day spent in a library would I live again, but rather some lurid day of labour and anxiety, of strife, or peril, or passion.

Occupied with this profound question, I scarcely noticed when my shade-sharer, with whom I sympathised only too keenly in her restless mood, rose and, lifting the light green curtain, passed out into the sunshine and was gone. Nor did I notice when the little wren ceased singing over-

---

head. At length recalled to myself I began to wonder at the unusual silence in the garden, until, casting my eyes on the lawn, I discovered the reason; for there, moving about in their various ways, most of the birds were collected in a loose miscellaneous flock, a kind of happy family. There were the starlings, returned from the fields, and looking like little speckled rooks; some sparrows, and a couple of robins hopping about in their wild startled manner; in strange contrast to these last appeared that little feathered clodhopper, the chaffinch, plodding over the turf as if he had hobnailed boots on his feet; last, but not least, came statuesque blackbirds and thrushes, moving, when they moved, like automata. They all appear to be finding something to eat; but I Watch the thrushes principally, for these are more at home on the moist earth than the others, and have keener senses, and seek for nobler game. I see one suddenly thrust his beak into the turf and draw from it a huge earthworm, a wriggling serpent, so long that although he holds his head high, a third of the pink cylindrical body still rests in its run. What will he do with it? We know how wandering Waterton treated the boa

---

which he courageously grasped by the tail as it retreated into the bushes. Naturally, it turned on him, and, lifting high its head, came swiftly towards his face with wide-open jaws; and at this supreme moment, without releasing his hold on its tail, with his free hand he snatched off his large felt hat and thrust it down the monster's throat, and so saved himself.

Just as I am intently watching to see how my hatless little Waterton will deal with *his* serpent, a startling bark, following by a canine shriek, then a yell, resound through the silent garden; and over the lawn rush those three demoniacal fox-terriers, Snap, Puzzy, and Babs, all determined to catch something. Away fly the birds, and though now high overhead, the baffled brutes continue wildly careering about the grounds, vexing the air with their frantic barkings. No more birds to-day! But now the peace-breakers have discovered me, and come tearing across the lawn, and on to the half-way chair, then to the hammock, scrambling over each other to inflict their unwelcome caresses on my hands and face.

Ah well, let them have their way and do their worst, since the birds are gone, and I shall go

---

soon. It is a consolation to think that they are not my pets; that I shall not grieve, like their mistress, when their brief barking period is over; that I care just so much and no more for them than for any other living creature, not excepting the *fer-de-lance,* "quoiled in the path like rope in a ship," or the broad-winged vulture "scaling the heavens by invisible stairs." None are out of place where Nature placed them, nor unbeautiful; none are unlovable, since their various qualities--the rage of the one and the gentleness of the other--are but harmonious lights and shades in the ever-changing living picture that is so perfect.

BIRDS IN A CORNISH VILLAGE

I

TAKING STOCK OF THE BIRDS

HAVING begun, or first written, this book in one village, which was near London, I am now finishing, or re-writing, it in another in "the westest part of all the land," over three hundred miles from the first. Here I had to go over this ancient work of twenty-three years ago, which was also my first English bird book, to prepare it for a new edition; and after all necessary corrections, omissions and additions of fresh matter made in the foregoing parts, it seemed best to throw out the whole of the concluding portion, which dealt mainly with the question of bird-preservation as

---

it presented itself at that time and is now out of date, thanks to the legislation of recent years and to the growth in this country of the feeling or desire for birds during the last two or three decades. In place of this discarded matter I propose to give here the results of recent observations on the bird life of a Cornish village.

My residence in the Cornish Village (or villages) was during May and June, 1915, and again from October of the same year to June, 1916. These were months of ill health, so that I was prevented from pursuing my customary outdoor rambling life; but, like that poor creature the barnyard fowl that can't use its wings, instinctively, or from old habit, I used my eyes in keeping a watch on the feathered (and flying) people about me.

The village, Lelant, is on the Hayle estuary, and to see the Atlantic one has but to walk past the grey old church at the end of the street, where the ground rises, to find oneself in a wilderness of towans, as the sand-hills are there called, clothed in their rough, grey-green marram grass and spreading on either hand round the bay of St. Ives. A beautiful sight, for the sea on a

---

sunny day is of that marvellous blue colour seen only in Cornwall; far out on a rock on the right hand stands the shining white Godrevy lighthouse, and on the left, on the opposite side of the bay, the little ancient fishing-town of St. Ives.

The river or estuary, in sight of the doors and windows of the village, was haunted every day by numbers of gulls and curlews. These last numbered about one hundred and fifty birds, and were always there except at full tide, when they would fly away to the fields and moors. Of all my bird neighbours I think that these gave me most pleasure, especially at night, when lying awake I would listen by the hour to the perpetual curlew conversation going on in the dark--an endless series of clear modulated notes and trills, with a beautiful expression of wildness and freedom, a reminder of lonely seashores and mountains and moorlands in the north country. What wonder that Stevenson, sick in his tropical island--sick for his cold grey home so many thousands of miles away, wished once more to hear the whaup crying over the graves of his forefathers, and to hear no more at all!

Of bird music by day there was little; you would

---

hear more of it in one morning in that small rustic village in Berkshire where the first part of this book was written than in a whole summer in one of these West Cornwall villages, so few comparatively are the songsters. Nor was this scarcity in the village only; it was everywhere, as I found when able to get out for a few hours during my two spring seasons in the place. Close by were the extensive woods of Trevalloe, where I was struck by the extraordinary silence and where I listened in vain for a single note from blackcap, garden-warbler, willow-wren, wood-wren, or redstart. The thrushes, chaffinch, chiff-chaff, and greenfinch were occasionally heard; outside the wood the buntings, chats, and the skylark were few and far between.

This scarcity of small birds is, I think, due in the first place to the extraordinary abundance of the jackdaw, the diligent seeker after small birds' nests, and to the autumn and winter pastime of bush-beating to which men and boys are given in these parts, and which the Cornish authorities refuse to suppress.

After a time, when, owing to increasing debility, I was confined more and more to the

---

BIRDS IN CORNISH VILLAGE 269

village, I began to concentrate my attention on a few common species that were always present, particularly on the three commonest--rook, daw, and starling; the first two residents, the starling, a winter visitor from September to April.

In October, I started feeding the birds at the house where I was staying as a guest, throwing the scraps on a lawn at the back which sloped down towards the estuary. First came all the small birds in the immediate neighbourhood--robin, dunnock, wagtail, chaffinch, throstle, blackbird, and blue and ox-eye tits. Then followed troops of starlings, and soon all the rooks and daws in the village began to see what was going on and come too, and this attracted the gulls from the estuary--I wished that it had drawn the curlews; and all these big ones were so greedy and bold, so noisy and formidable-looking that the small birds were quite driven out; all except the starlings that came in hungry crowds and were determined to get their share.

At the beginning of December I had to move to a nursing-home at the Convent of the Sisters of the Cross at the adjacent village of Hayle, just across the estuary. The Convent buildings and

---

grounds and gardens are fortunately outside the ugly village, and my room had an exceptionally big window occupying almost the whole wall on one side, with an outlook to the south over the green fields and moors towards Helston. An ideal sick-room for a man who can't be happy without the company of birds, and here, even when lying on my bed before I was able to sit or stand by the window, a large portion of the sky, rainy or blue, was visible, and rooks and daws and gulls and troops of starlings, and the curlews from the river, were seen coming and going all day long.

But it was much better when I was able to go to the window, since now, by feeding them, I could draw the birds to me. I fed them on a green field beneath my window, where the Convent milch-cows were accustomed to graze for some hours each day. All through the winter there was grass for them, and I was glad to have them there, as the cow is my favourite beast, and it was also pleasant to see the wintering starlings consorting with them, clustering about their noses, just as they do in the pasture lands in summer time. But I found it best to feed the birds when

---

the cows were not there, on account of the behaviour of one of them, a young animal who had not yet been sobered by having a calf of her own. She was a frivolous young thing and when tired of feeding, she would start teasing the old cows, pushing them with her horns, then flinging up her hind legs to challenge them to a romp. The sight of a crowd of birds under my window would bring her at a gallop to the spot to find out what all the fuss was about, and the birds would be driven off.

One morning I was at my window when the field was empty of bird and beast life with the exception of a solitary old rook, a big bird who was a constant attendant and so much bigger than most of the rooks that I had come to know it well. By and by the young cow walked into the field by herself and, after gazing all round as if surprised at finding the place so lifeless, she caught sight of and fixed her eyes on the old rook working at the turf some fifty or sixty yards away. Presently she began walking towards it, and when within about twenty yards put her head down and charged it. The rook paid no attention until she was almost on it, then rose up,

---

emitting its angriest, most raucous screams while hovering just over her head, and having thus relieved its indignant feelings it flew heavily away to the far end of the field, and settling down began prodding away at the soil. The cow, standing still, gazed after it, and one could almost imagine her saying: "So you won't get out of the field! Well! I'll soon make you. I'm going to have it all to myself this morning." And at once she began rapidly walking towards the bird. But half-way to it was the post set up in the middle of the field for the cows to rub their hides, and on coming abreast of it the sight of it and its proximity suggested the delight of a rub, and turning off at right angles she walked straight to the post and began rubbing herself against it. The rook went on with its business, and after that there was no more quarrelling.

Another morning this same old rook came with his mate to the field: separating, they came down a distance of a hundred yards or more apart and began searching for grubs. By and by the old cock discovered something particularly good and after vigorously prodding the turf for a few moments he sprang up and flew excitedly to his

---

mate, who instantly knew what this action meant and began fluttering her wings and crying for the dainty morsel which he proceeded to deliver into her wide-open mouth. Having fed her, he flew back to the same spot and began working again.

This is a common action of the rooks, and I saw this same bird feed his mate on other occasions during the winter months, when I have no doubt that he, poor wretch, could hardly find food enough to keep himself alive during the dark season of everlasting wind and rain when the dim daylight lasted for about six hours. But I never saw a daw or starling feed his mate, or feed another daw or starling, although I watched closely every day and often for an hour at a stretch, and though I am convinced that the starling, like the rook and crow and daw, and in fact all the Corvidae, pairs for life. To this point I will return presently; let me first relate another incident about our frivolous and irresponsible young cow.

One morning when the cows were in the field, some herring-gulls drifted by and a few of them remained circling about above the field. I threw

---

out a piece of bread, and a troop of starlings rushed to it, and one of the gulls dropped down and took possession of it, but had scarcely began tearing at it when two more gulls dropped down and the first bird, lifting his wings began screaming "Hands off!" at the others, and the others, also raising their wings, screamed their wailing screams in reply. The young cow, attracted by the noise, gazed at them for a few moments, then all at once putting her head down furiously charged them. The three gulls rose up simultaneously and floated over her and then away, leaving her standing on the spot, shaking her head in anger and disgust at their escape. A rhinoceros charging a ball of thistledown or a soap-bubble, and causing it to float away with the wind it created,

would not have been a more ludicrous spectacle.

II

DO STARLINGS PAIR FOR LIFE?

FROM my boyhood, when I first began to observe birds, I started with the imbibed notion that those which paired for life were the rare exceptions--the dove that rhymed with love, the eagle, and perhaps half a dozen more. Who, for instance, would imagine that the sexes could be faithful in parasitical species like the cuckoo of Europe and the cow-birds of America? Yet even as a boy I made the discovery that an Argentine cow-bird that lays its eggs in the nests of other species, does actually pair for life; and so effectually mated is it, that on no day and no season of the year will you see a male without his female: if he flies she flies with him and feeds and drinks with him, and when he perches she perches at his side, and he never utters a sound but a responsive sound immediately falls from her devoted beak.

Again, it may seem unlikely that there can be pairing for life in species, like the chaffinch of northern Europe and, with us, of Scotland, in which the sexes separate and migrate separately. Also of non-gregarious species like the nightingale in which the males arrive in this country several days before the females. Yet I am confident that if we could catch and mark a considerable number of pairs it would be found that the same male and female found one another and remated every year.

It comes to this, that birds may pair for life, yet not be all the time or all the year together, as in the case of hawks, crows, owls, herons, and many others. In numberless species which undoubtedly pair for life the sexes keep apart during several hours each day, and there is some evidence that those that separate for a part of the year remain faithful.

An incident, related by Miss Ethel Williams, of Winchester, in her natural history notes contributed to a journal in that city, bears on this point. She had among the bird pensioners in the garden of her house adjoining the Cathedral green, a female thrush that grew tame enough to

BIRDS IN CORNISH VILLAGE 277

fly into the house and feed on the dining-room table. Her thrush paired and bred for several seasons in the garden, and the young, too, were tame and would follow their mother into the house to be fed. The male was wild and too shy ever to venture in. She noticed the first year that it had a wing-feather which stuck out, owing probably to a malformation of the socket. Each year after the breeding season the male vanished, the female remaining alone through the winter months, but in spring the male came back--the same bird with the unmistakable projecting wing-feather. Yet it was certain that this bird had gone quite away, otherwise he would have returned to the garden, where there was food in abundance during the spells of frosty weather. As he did not appear it is probable that he migrated each autumn to some warmer climate beyond the sea.

I have noticed that wagtails, thrushes, blackbirds, and some other species when the young are out of the nest, divide the brood between male and female and go different ways and spend the daylight hours at a distance apart, each attending to the one or two young birds in its charge.

278 BIRDS IN TOWN AND VILLAGE

One winter, a few years ago, I was staying for a few days at a cottage facing Silchester Common, and on going out after breakfast to feed the birds I particularly noticed a male grey wagtail among those that came to me, on account of its beauty and tameness. Every morning I fed it, and on my speaking to my landlady about it she said, "Oh, we know that bird well; this is the fourth winter it has spent with us, but it always came before with its mate. The poor little thing had only one leg, but managed to hop about and feed very well; this year the poor thing didn't turn up with its mate, so we suppose it had met its death somewhere during the summer."

I have often watched the gatherings of pied wagtails (always with a certain number of the grey species among them) in places where they spend the winter in our southern counties, at some spot where they are accustomed to congregate each evening to hold a sort of frolic before going to roost, and it has always appeared to me that the birds, both pied and grey, were in pairs. So too, in watching the starlings day after day in the field in front of my window. Well able with my binocular to observe them closely, I saw much

---

to convince me that the starling, too, lives all the year with his mate.

Each morning the birds that had made our village their daily feeding-ground, would, on arrival from the roosting-place in one body, break up into numerous small parties of half a dozen to twenty or more birds. All day long these little flocks were hurrying about from field to field, spending but a short time at one spot, so hungry were they and anxious to find a more productive one, and in every field they would meet and mix with other small groups, and presently all would fly, and breaking up into small parties again go off in different directions. Thus one had a constant succession of little flocks in the field from morning till night, and I found from counting the birds in each small group that in three cases in four they were in even numbers. Again, I have often seen a group of three, five, seven or nine birds on the field, and after a while a solitary starling from a neighbouring field or from some treetop near by has flown down to join the group and make the numbers even.

The birds when feeding, I have said, are always in a desperate hurry, and little wonder, since

---

after a night, usually wet and cold, of from sixteen to eighteen hours and only about six to feed in, they must be in a half-starved state and frantic to find something to swallow. No sooner do they alight than they begin running about, prodding with their beaks, and all the time advancing, the birds keeping pretty well abreast. Now, from time to time you will notice that a bird finds something to delay him and is left behind by the others. On they go--prod, prod, then a little run, then prod, prod again and run again--while he, excited over his find, and vigorously digging at the roots of the grass, lets them go on without him until he is yards behind. Whenever this happens you will see one of the advancing birds pause in its prodding to look back from time to time as if anxious about the one left behind; and by and by this same bird, its anxiety increasing, will suddenly spring into the air and fly back to place itself at the side of the other, to wait quietly until it has finished its task; and no sooner does the busy one put up its head to signal that he is ready than up they spring and fly together on to the flock. No one witnessing this action can doubt for a moment that these two are mates,

---

and that wherever they paired and bred originally

--in Lincoln or York or Thurso or perhaps in one of the western islands--they paired for life and will stick together, summer and winter and in all their wanderings, as long as they live.

Until one observes starlings in this close way, even to their minutest actions--I had indeed little else to do during my three winter months in this nursing-home--it is only natural to believe that among gregarious species the starling is one of those least likely to pair for life, seeing that in it the gregarious instinct is intensified and more highly developed than in most others. One would suppose that the flock, which is like an organism

--that is to say, the attachment to the flock

--would, out of the breeding season, take the place of the close relation or companionship between bird and bird seen in species known to pair for life. Only the pairing passion, one would suppose, could serve to dissolve the company of birds and this only for a brief season of about a couple of months' duration. There is but one brood raised in the season, and the whole business of reproduction is well over before the end of June. Later breeders are those that have lost

their first eggs or broods. And no sooner are the young brought off and instructed in the starling's sole vocation (except his fruit-eating) of extracting the grubs it subsists on from the roots of the grass--a business which detains them for a week or two--than the married life is apparently over and the communal life resumed. The whole life of the bird is then changed; the sole tie appears to be that of the flock; home and young are forgotten: the birds range hither and thither about the land, and by and by migrate to distant places, some passing oversea, while others from the northern counties and from Scotland and the islands come down to the south of England, where they winter in millions and myriads. There they form the winter habit of congregating in immense numbers in the evening at their favourite roosting-places, and hundreds and thousands of small flocks, which during the daylight hours exist distributed over an area of hundreds of square miles all make to one point and combine into one flock. At such times they actually appear to rejoice in their own incalculable numbers and gather earlier than they need at the roosting-place, so that the whole vast gathering

may spend an hour or so in their beloved aerial exercises.

To anyone who witnesses these gatherings and sees the birds rising from time to time from the wood, and appearing like a big black cloud in the sky, growing lighter and darker alternately as the birds scatter wide or mass themselves in a closer formation, until after wheeling about for some minutes they pour back into the trees; and who listens to the noise they make, as of a high wind in the wood, composed, as it is, of an infinity of individual voices, it must seem incredible that all these birds can keep in pairs. For how could any couple hold together in such circumstances, or when separated ever meet again in such a multitude, or, should they ever meet by chance, how recognize one another when all are exactly alike in size, shape, colour and voice?

They can, and certainly do, keep together, and when forced apart as, when pursued by a hawk, they scatter in all directions, they can quickly find one another again. They can do it because of their perfect discipline, or instinct, or the perfection of the system they follow during their autumn and winter wanderings and migrations.

The breeding season over, the birds in each locality unite in a small flock composed of twenty or thirty to fifty or more pairs and start their wandering life. Those in the north migrate or drift south, and vast numbers, as we see, spend the winter in the southern counties. And here they have their favourite roosting-places and are accustomed to assemble in tens and hundreds of thousands. But the original small flock composed of a few pairs, is never broken up--never absorbed by the multitude. Each morning when it is light enough, the birds quit the roosting-wood, but not all together; they quit it in flocks, flock following flock so closely as to appear like a continuous stream of birds, and the streams flow out in different directions over the surrounding country. Each stream of birds is composed of scores and hundreds of units, and each unit drops out of the stream and slopes away to this or that side, to drop down on its own chosen feeding-ground, to which it returns morning after morning through the winter. When all the units have dropped out and settled on their feeding areas for the day, it may be seen that the whole country within a circuit of ten or twelve or more miles from the roosting-

---

place has been occupied, that each flock has its own territory, where it splits up into some groups and spends its short hours flying about and exploring every green field, and one might almost say "every grass." One can only explain this perfect distribution by assuming that each unit instinctively looks for unoccupied ground in its winter habitat, and that consequently there is very little overlapping. It must also be assumed that at the place of assembly in the evening each flock has its own roosting-place--its own trees and bushes where the members of the flock can still keep together and to which after each aerial performance they can return. The flock comes back to sleep on its own tree, and no doubt every couple roosts side by side on its own twig.

On the return of Spring the birds do not migrate in a body, but slip away, flock by flock, to reappear about the end of April in their old breeding-place in the North Country, with, perhaps, the loss of a few members--the one that was old and died in the season of scarcity; and one that was taken at the roost by a brown owl, and one that had its feet frozen to the perch; and was killed by a jackdaw when struggling to free

itself; and one that was struck down by a sparrowhawk on his homeward journey.

What I have so far been unable to trace is the career of the young after August. We see that once they are able to fend for themselves they club together in small flocks and continue together during their "brown thrush" stage, but by and by they get the adult plumage and language and are no longer distinguishable as young. Do they, then, join the old birds before the wandering and migrating south begins? And do they

pair or not before the winter?

III

VILLAGE BIRDS IN WINTER

THROUGHOUT the winter of 1915-16, and more particularly during my three months in the hospital at Hayle, from the beginning of December to March, I was greatly impressed at the perpetual state of hunger in which the birds exist, especially the three commonest species in our village--rook, daw, and starling. Little wonder that the sight of a piece of bread thrown out on the green field below my window would bring all these three and many others with a rush from all sides, every one eager to get a morsel! But the birds that live most in a groove, as it were, like the rook and starling, and have but one kind of food and one way of finding it, are always the worst off in winter. These subsist on the grubs and other minute organisms they are able to pick out of the grass roots, and are life workers paid by the

piece who must labour hard and incessantly to make enough to keep themselves alive; their winter life is accordingly in startling contrast to that of the daw--one that lives on his wits and fares better and altogether has an easier and more amusing time.

It was the habit of the three species named to quit the wood where they roosted as soon as it was light enough for them to feed, the time varying according to the state of the weather from half-past eight to ten o'clock, the mornings being usually wet and dark. The rooks that had their rookery in the village numbered forty or fifty birds, and these would remain at the village, getting their food in the surrounding fields for the rest of the day. The daws would appear in a body of two or three hundred birds, but after a little while many of them would go on to their own villages further away, leaving about sixty to eighty birds belonging to the village. Last of all the starlings would appear in flocks and continuous streams of birds often fighting their way against wind and rain, leaving about a couple of hundred or more behind, these being the birds that had settled in the village for the season, and

worked in the grass fields in and surrounding it. Rooks and starlings would immediately fall to work, while the daws, the flock breaking up into small parties of three or four, would distribute themselves about the village and perch on the chimneypots. They would perch and then fly, and for all the rest of the day would be incessantly shifting about from place to place, on the look-out for something to eat, dropping from time to time to snatch up a crust of bread or the core of an apple thrown away by a child in the road, or into a back garden or on to a dust-heap where potato-parings and the head of a mackerel or other refuse had been thrown. They were very bold, but not as courageous as the old-time British kite that often swooped to snatch the bread from a child's hand.

From time to time one, or a pair, of a small party of these daws would drop down on the field before my window when the rooks and starlings were there prodding busily at the turf, but though I watched them a thousand times I never detected them trying to find something for themselves. They simply stood or walked about among the working birds, watching them intently. Grub-

finding was an art they had not acquired, or were too indolent or proud to practise; but they were not too proud to beg or steal; they simply watched the other birds in the hope of being able to snatch up a big unearthed grub and run away with it. As a rule after a minute or two they would get tired of waiting and rush off with a lively shout. Back they would go to the chimneypots and to their flying up and down, suspending their flight over this or that yard or garden, and by and by one would succeed in picking up something big, and at once all the other daws in sight would give chase to take it from him; for these village daws are not only parasites and cadgers, but worse--they are thieves without honour among themselves.

In spite of all the time and energy wasted in their perpetual races and chases going on all over the village, every bird exerting himself to the utmost to rob all he can from his pals, they get enough to eat; for when the day is over and other daws from other villages drop in to visit them, all unite in a big crowd and wheel about, making the place ring with their merry yelping cries, before sailing away to the wood. One

might say after witnessing and listening to this evening performance that they have great joy in their rascally lives.

But for the poor starling there is little joy in these brief, dark, wet winter days, even if there is little frost in this West Cornwall climate. A frost of a few days' duration would be fatal to incalculable numbers, especially if, as in the great frosts of the winters of 1894-5 and 1896-7, severest in the south and west of England, it should come late in winter, I think it can be taken as a fact that a long or overseas migration takes place before midwinter or not at all. In January and February, when birds are driven to the limits of the land by a great cold they do not cross the sea, either because they are too weak to attempt such an adventure or for some other reason unknown to us. We see that on these occasions they come to the seashore and follow it south and west even to the western extremity of Cornwall, and then either turn back inland or wait where they are for open weather, many perishing in the meantime.

During those three winter months, when I watched the starlings at work on the field before my hospital window, they appeared to be in a perpetual state of extreme hunger and were always running over the ground, rapidly prodding as they moved, and apparently finding their food almost exclusively on the surface--that is to say, on the surface of the soil but under the grass, at its surface roots. At other seasons they go deep when they know from the appearance of every blade of grass whether or not there is a grub feeding on its roots beneath the surface. Without shooting and examining the stomachs of a large number of starlings it was not possible to know just what the food consisted of; but with my strong binocular on them I could make out that at almost every dig of the beak something was picked up, and could actually see it when the beak was held up with the minute morsel at its tip--a small, thread-like, semi-transparent worm or grub in most instances. Two or three of these atomies would hardly have made a square meal for a ladybird, and I should think that a starling after swallowing a thousand would fed very hungry. And on many days this scanty, watery food had to be searched for in very painful conditions, as it rained heavily on most days and

often all day long. At such times the birds in their sodden plumage looked like drowned starlings fished out of a pool and galvanized into activity. Nor were they even seen to shake the wet off--a common action in swallows and other birds that feed in the rain; they were too hungry, too anxious to find something to eat to keep the starling soul and body together before the long night of eighteen or twenty hours would overtake them.

No doubt the winter of 1915-16 was exceptionally wet and cold, although without any severe frosts; a long frost in February, when the birds were most reduced, would probably have proved fatal to at least half their number. But though it continued wet and cold, things began to mend for the starlings towards the end of February, and in March the improvement was very marked; they were not in such a perpetual hurry; their time was longer now, and by the end of the month their working day had increased from five or six to twelve or fourteen hours, and the light had increased and grubs were easier to find. By April, the starlings no longer appeared to be the same species as the poor, rusty, be-

draggled wretches we had been accustomed to see; they are now lively, happy birds with a splendid gloss on their feathers and beaks as bright a yellow as the blackbird's. Finally, in April they left us, not going in a body, but flock by flock, day after day, until by the end of the month all were gone back to their homes in the north--all but the two or three to half a dozen pairs in each village. And these few that stay behind are

new colonists in West Cornwall.

IV

INCREASING BIRDS IN BRITAIN

ABOUT the daw, or Jackie, or Dorrie or Jackie-Dorrie, as he is variously and familiarly called, and his village habits, there will be more to say presently; just now my concern is with another matter--a veritable daw problem.

For the last twenty years or longer it has seemed to me that the daw is an increasing species in Britain; at all events I am quite sure that it is so in the southern half of England, particularly along the coast of Somerset, Devon, Dorset, and in Cornwall, more than in any other county. And why is it? He is certainly not a respectable bird, like the starling, for example--if we do not go to the cherry-grower for the starling's character. He is and always has been on the keeper's and farmer's black list, and scarcely a week passes but you will find him described in some gamekeeper's

---

or farmer's journal as "even worse than the rook." Even the ornithologists who are interested in birds as birds haven't a good word to say of the daw. According to them he alone is responsible for the disappearance of his distinguished relation, the chough. (The vulgar daw is of course devoid of any distinction at all, unless it be his grey pate and wicked little grey eyes.)

The ornithologists were wrong about the chough, just as they had been wrong about the goldfinch, during the late years of the nineteenth century, and as they were wrong about the swallows and martins in later years. Of the goldfinch, they said, and solemnly put it down in their books, that owing to improved methods of agriculture the thistle had been extirpated and the bird, deprived of his natural food, had forsaken this country. But no sooner did our County Councils begin to avail themselves of the powers given them by the Bird Act of twenty years ago to protect the goldfinch from the bird-catcher, than it began to increase again and is still increasing, year by year, all over the country.

Of the decrease of swallows and martins, they

said it resulted from the action of the sparrows in ousting them from their nests and nesting-sites. But we know the true cause of the decline of these two species, the best loved and best protected of all birds in Britain, not even excepting robin redbreast. The French Government, in response to representations on this matter from our Foreign Office, have caused enquiries to be made and have found that our swallows are being destroyed wholesale in France during the autumn migration, and have promised to put a stop to this deplorable business. They do not appear to have done so, since the promise was made three years ago, and I can say from my own observation in the south and west countries that the decline has continued and that we have never had so few swallows come to us as in the present summer of 1916.

The daw--to return to that subject--has always been regarded as an injurious species, and down to a quarter of a century ago every farm lad in possession of a gun shot it in the interests of the henwife, even as he had formerly shot the kite, a common British species and a familiar feature in the landscape down to the early years

of last century. Doubtless it was a great thing to bring down this great bird "that soars sublime" and nail it to the barn-door. By the middle of the last century it had become a rarity, and the ensuing rush for specimens and eggs for private collectors quickly brought about its virtual extinction. The kite is but one of several species--six of them hawks--extirpated within the last forty years. Why, then, does the daw, more injurious to the game-preserver and henwife than any one of these lost hawks, continue to flourish and increase in numbers? It is, I imagine, because of the growth of a sentiment which favours its preservation. But it is not the same as that which has served to preserve the rook and made it so common. That is a sentiment confined to the landowning class--to those who inherit great houses where the ancient rookery with its crowd of big, black, contentious birds caw-cawing on the windy elms, has come to be an essential part of the establishment, like the gardens and park and stables and home-farm and, one might add, the church and village. This sentiment differs, too, from the heron-sentiment, which serves to keep that bird with us in spite of the annual wail, ris-

---

ing occasionally in South Devon to a howl, of human trout-fishers. It is a traditional feeling coming down from the far past in England--from the time of William the Conqueror to that of William of Orange and the decay of falconry. That a species without any sentiment to favour it and without special protection by law may increase is to be seen in the case of the starling. This increase has come about automatically after we had destroyed the starling's natural enemies and then ceased to persecute it ourselves. Of all birds it was the most preyed on by certain raptorial species, especially by the sparrowhawk, which is now becoming so rare, assisted by the hobby (rarer still) and the merlin. It was more exposed than other birds to these enemies owing to its gregarious and feeding habits in grasslands and the open country, also to its slower flight. The greatest drain on the species, came, however, from man. The starling was a favourite bird for shooting-matches up till about thirty years ago, and was taken annually in large numbers by the bird-catchers for the purpose. It is probable that this use of the bird for sport caused people to eat it, and so common did the habit become that

---

at the end of summer, or before the end, shooting starlings for the pot was practised everywhere. Old men in the country have told me that forty or fifty years ago it was common to hear people on the farms say that of all birds the starling was the best to eat.

When starling and sparrow shooting-matches declined, the starling went out of favour as a table-bird, and from that time the species has been increasing. At present the rate of increase grows from year to year, and during the last decade the birds have colonized every portion of the north of Scotland and the islands, where the starling had previously been a rare visitor--a bird unknown to the people. Here in West Cornwall where I am writing this chapter the starling was only a winter visitor until recently. Eight years ago I could only find two pairs breeding in the villages--about twenty-five in number--in which I looked for them; in the summer of 1915 I found them breeding in every town and village I visited. At present, June, 1916, there are six pairs in the village I am staying at. It may be the case, and from conversations I have had with farmers about the bird I am inclined to believe

---

it is so, that a strong feeling in favour of the starling (in the pastoral districts) is growing up at the present time, a feeling which in the end is more powerful to protect than any law; but such a feeling has not become general as yet, and consequently has had nothing to do with the extraordinary increase of the bird.

The wood-pigeon is another species which, like the starling, has increased greatly in recent years, without special protection and with no sentiment in its favour. . . . The sentiment is all confined to the nature-lovers, whose words have no effect on the people generally, least of all on the farmers. I am reminded here of the experience of a young man, an ardent bird-lover, on his visit to a Yorkshire farm. His host, who was also a young man, took him a walk across his fields. It was a spring day of brilliant sunshine, and the air was full of the music of scores of soaring skylarks. The visitor long in cities pent, was exhilarated by the strains and kept on making exclamations of rapturous delight, "Just listen to the larks! Did you ever hear anything like it!" and so on.

His host, his eyes cast down, trudged on in

---

302 BIRDS IN TOWN AND VILLAGE

glum silence. Finally the young man, carried away by his enthusiasm, stopped and turning to his companion shouted, "Listen! Listen! Do you hear the larks?"

"Oh, yes," drawled the other, looking more glum than ever, "I hear them fast enough. And I wish they were all dead!"

So with the other charming species. The moan of doves in immemorial elms is a pleasing sound to the poets, but it does not prevent the farmers throughout the land from wishing them all dead; and every person who possesses a gun is glad to help in their massacre. For the bird is a pest and he who shoots it is doing something for England; furthermore, shooting it is first-rate sport, not like slaughtering wretched little sparrows or innocent young rooks just out of their windy cradles. And when shot it is a good table-bird, with as much tasty flesh on it as a woodcock or partridge.

How, then can we account for the increase of such a species? One cause is undoubtedly to be found in the removal by gamekeepers of its three chief enemies--the carrion crow, magpie, and jay--all these three being great devourers of pigeon's eggs, which of all eggs are most con-

---

BIRDS IN CORNISH VILLAGE 303

spicuous and open to attack. Then again the winter immigration of wood-pigeons from northern Europe appears to be on the increase, and it may be conjectured that a considerable number of these visitors remain annually to breed with us. There has also been an increase in the stockdove and turtledove in recent years, and the former species is extending its range in the north. The cause or causes of the increase of the turtledove are not far to seek. Its chief feathered enemies, the egg and fledgling robbers, are the same as the wood-pigeon's; moreover, the turtledove is least persecuted by man of our four pigeons, and being strictly migratory it quits the country before shooting-time begins; add to this that the turtledove has been specially protected under Sir Herbert Maxwell's Act of 1894 in a good number of English counties, from Surrey to Yorkshire.

Of the stockdove we can only say that, like the ring-dove, it has increased in spite of the persecution it is subject to, since no person out after pigeons would spare it because it is without a white collar. With the exception of the county of Buckinghamshire it is not on the schedule any-

where in the country. One can only suppose that this species has been indirectly benefited by the bird legislation and all that has been done to promote a feeling favourable to bird-preservation during the last thirty

years.

V

THE DAW SENTIMENT

I HAVE spoken of the wood adjacent to the villages of Hayle and Lelant where the rooks, daws, and starlings of the neighbourhood have their winter roosting-place. This is at Trevelloe, the ancient estate of the Praeds, who now call themselves Tyringham. Here the daws congregate each evening in such numbers that a stranger to the district and to the local habits of the bird might imagine that all the cliff-breeding jackdaws in West Cornwall had come to roost at that spot. Yet the cliff-breeders, albeit abundant enough, are but a minority of the daw population of this district. The majority of these birds live and breed in the neighbouring villages and hamlets--St. Ives, Carbis Bay, Towadneck, Lelant, Phillack, Hayle, and others further away. It is a jackdaw metropolis and, as we have seen, every village receives its own quota of birds each morning, and

---

there they spend the daylight hours and subsist on the waste food and on what they can steal, just as the semi-domestic raven and the kite did in former ages, from Roman times down to the seventeenth century.

Early in May the winter congregation breaks up, the cliff-breeders going back to the rocks and the village birds to their chimneys, where they presently set about relining their old nests. There are plenty of places for all, since there are chimneys in almost every cottage where fires are never lighted, and as ventilation is not wanted in bedrooms the birds are allowed to bring in more materials each year, until the whole flue is filled up. Year by year the materials brought in, sink lower and lower until they rest on the closed iron register and change in time to a solid brown mould. Thus, however long-lived a daw may be--and there are probably more centenarians among the daws than among the human inhabitants of the villages--it is a rare thing for one to be disturbed in his tenancy.

In the cottage opposite the one I was staying in, its owner, an old woman who had lived in it all her life, had recently died, aged eighty-seven.

She was very feeble at the last, and one cold day when she could not leave her bed, the extraordinary idea occurred to some one of her people that it might be a good thing to light a fire in her room. The fireplace was examined and was found to have no flue, or that the flue had been filled with earth or cement. The village builder was called in, and with the aid of a man on the roof and poles and various implements he succeeded in extracting two or three barrow-loads of hard earth which had no doubt once been sticks, centuries ago, as the building was very ancient. No one had remembered that the daws had always occupied the same chimney; the old dame herself had seen them going in and out of it from her childhood, and her end was probably hastened by the disturbance made in cleaning it. Now she is gone the daws here are in possession of it once more.

All through the month of May daws were to be seen about the village, dropping from time to time upon the chimneypots where they had their nests and occasionally bringing some slight materials to form a new lining, but it was very rare to see one with a stick in his beak. The flues

---

were already full of old sticks and no more were wanted. It was amusing to see a bird flying about, suddenly tumble out of the air on to a chimneypot, then with tail tipped up and wings closed, dive into the cavity below. One wondered how the young birds would be got out!

Talking with the rector of the neighbouring parish of Phillack one day on this subject, he said, "Don't imagine that the daws restrict themselves to the chimneys where fires are not lighted. At all events it isn't so at

Phillack. Perhaps we have too many daws in our village, but every year before lighting fires in the drawing and dining-rooms we have to call in a man with a pole to clear the flues out." He told me that a few years ago, one cold June day, a fire was lighted in the drawing-room, and as the smoke all poured out into the room a man was sent up to the roof with a pole to clear the obstruction out. Presently a mess of sticks came down and with them two fully-fledged young jackdaws, one dead, killed with the pole, the other sound and lively. This one they kept and it soon became quite tame; when able to fly it would go off and associate with the wild birds, but refused to leave

---

the house until the following summer, when it found a mate and went away.

The head keeper at Trevelloe, a remarkably vigorous and intelligent octogenarian who has been in his place over half a century, gave me some interesting information about the daws. He says they have greatly increased in recent years in this part of Cornwall because they are no longer molested; no person, he says, not even a gamekeeper anxious about his pheasants, would think of shooting a jackdaw. But this is not because the bird has changed its habits. He is as great a pest as ever he was, and as an example of how bad jackdaws can be, he related the following incident told him by a friend of his, a head keeper on an estate adjoining a shooting his master took one year on the northwest coast of England. It happened that a big colony of daws existed within a mile or two of the preserves, and one day the keeper was called' away in a hurry and left the coops unattended for the best part of a day; it was the biggest mistake he had ever made and the chief disaster of his life. On his return he found that the daws had been before him and that all his precious chicks had been car-

---

ried off. For several hours of that day there was a steady coming and going of birds between the cliffs and the coops, every daw going back with a chick in his beak for his hungry young in the nest.

Yet my informant, this ancient and singularly intelligent old man, a gamekeeper all his life, who knows his jackdaw, could not tell me why gamekeepers no longer persecute so injurious a bird I He will not allow a sparrowhawk to exist in his woods, yet all he could say when I repeated my question was, "No keeper ever thinks of hurting a jack now, but I can't say why."

The reason of it I fancy is plain enough; it is simply the sentiment I have spoken of. In a small way it has always existed in certain places, in towns, where the jackdaw is associated in our minds with cathedrals and church towers--where he is the "ecclesiastical daw"; but the modern wider toleration is due to the character, the personality, of the bird itself, which is more or less like that of all the members of the corvine family, with the exception of the rook, who always tries his best to be an honest, useful citizen; but it is not precisely the same. They may be regarded

as bad hats generally In the bird community, and on this very account--"I'm sorry to say," to quote Mr. Pecksniff--they touch a chord in us; and the daw being the genial rascal in feathers par excellence is naturally the best loved.

It has thus come about that of all the Corvidae the daw is now the favourite as a pet bird, and in the domestic condition he is accorded more liberty than is given to other species. We think he makes better use of his freedom, that he does not lose touch with his human friends when allowed to fly about, and appears more capable of affection.

Formerly, the raven and magpie came first as pets. The raven vanished as a pet, because like the goshawk, kite, and buzzard, he was extirpated in the interests of the game-preserver and henwife. The magpie was then first, and has only been recently ousted from that ancient, honourable position. The pie was a superior bird as a feathered pet in a cage; he is beautiful in shape and colour in his snow-white and metallic dark-green and purple-glossed plumage, and his long graduated tail. Moreover, he is a clever bird. To my mind there is no more fascinating species

---

when I can find it in numbers, in places where it is not persecuted, and is accustomed to congregate at intervals, not as rooks and starlings do merely because they are gregarious, but purely for social purposes--to play and converse with one another. Its language at such times is so various as to be a surprise and delight to the listener; while its ways of amusing itself, its clowning and the little tricks and practical jokes the birds are continually playing on each other, are a delight to witness. All this is lost in a caged bird. He is handsome to look at and remarkably intelligent, but he distinguishes between magpies and men; he doesn't reveal himself; his accomplishments, vocal and mental, are for his own tribe. In this he differs from the daw; for the daw is less specialized; he is an undersized common crow, livelier, more impish than that bird, also more plastic, more adaptive, and takes more kindly to the domestic or parasitic life. Human beings to him are simply larger daws, and unlike the pie he can play his tricks and be himself among them as freely as when with his feathered comrades. We like him best because he makes himself one of us.

Undoubtedly the chough comes nearest to the daw mentally, and as it is a far more beautiful bird--the poor daw having little of that quality--it would probably have been our prime favourite among the crows but for its rarity. Formerly it was a common pet bird, caged or free, in all the coast districts where it inhabited, and it may be that the desire for a pet chough was the cause of its decline and final disappearance all round the south and west coasts of England, except at one spot near Tintagel where half a dozen pairs still exist only because watchers appointed by the

Royal Society for the Protection of Birds are always on the spot to warn off the nest-robbers during the breeding season. But of the chough in captivity or as a domesticated bird we know little now, as no records have been preserved. I have only known one bird, taken from a North Devon cliff about forty years ago, at a house near the coast; a very beautiful pet bird with charming, affectionate ways, always free to range about the country and the cliffs, where it associated with the daws. It was the last of its kind at that place, and I do not know if it still lives.

Next to the chough the jay comes nearest to

---

the daw mentally of all our crows, and as he excels most of our wild birds in beauty he would naturally have been a first favourite as a pet but for the fact that it is only in a state of nature in which he is like the daw--lively, clever, impish; in captivity he is more like the magpie and affiliates even less than that bird with his human associates. In confinement he is a quiet, almost sedate, certainly a silent bird: He is essentially a woodland species; all his graces, his various, often musical, language, with many imitations of bird and animal sounds, and his spectacular games and pretty wing displays, are for his own people exclusively. He must have his liberty in the woods and a company of his fellow-jays to exhibit his full lustre.

The difference between jay and daw is similar to that between fox and dog; or rather let us say, between one of the small desert foxes of Syria and Egypt--the fennec, for instance--and the jackal, the domestic dog's progenitor; the first gifted with exquisite grace and beauty, was too highly specialized to suit the domestic condition; hence the generalized unbeautiful beast was chosen to be man's servant and companion. In

---

the same way it looks as if we were taking to the daw in preference to the more beautiful bird because he is more like us, or understands us better, or adapts himself more readily to our way of life.

I believe that about nine out of every ten interesting and amusing stories about charming pet birds I have heard in England during the last quarter of a century relate to the daw, and this, I think, goes to show that he is a prime favourite as a feathered pet, at all events in the southern and western counties.

VI

STORY OF A JACKDAW

WHEN I laid my pen down after concluding Part V it pleased me to think that I had written the last word, that, my task finished, I was free to go on to something else. But I was not yet wholly free of the jackdaws; their yelping cries were still ringing in my mental ears, and their remembered shapes were still all about me in their black dress, or cassock, grey hood, and malicious little grey eyes. The persistent images suggested that my task was not properly finished after all, that it would be better to conclude with one of those anecdotes or stories of the domesticated bird which I have said are so common; also that this should be a typical story, which would serve to illustrate the peculiar daw sentiment--the affectionate interest we take in him, not only in spite of his impudence and impishness and naughtiness, but also to some extent because of these same

---

qualities, which find an echo in us. Accordingly I set myself to recall some of the latest anecdotes of this kind which I had heard, and selected the one which follows, not because it was more interesting as a daw story than the others, but mainly on account of the shrewd and humorous and dramatic way in which it was related to me by a little boy of the working class.

I met him on a bright Sunday morning at the end of June in the park-like grounds of Walmer Castle. I had not long been seated on a garden bench when a daw came flying to a tree close by and began craning her neck and eyeing me with one eye, then the other, with an intense, almost painful curiosity; and these nervous movements and gestures immediately revealed to me that she had a nestful of young birds somewhere close by. After changing her position several times to view me from other points and find out what I was there for, she came to the conclusion that I was not to be got rid of, and making a sudden dash to a tree standing just before me, disappeared in a small hole or cleft in the trunk about forty-five feet above the ground, and in a few seconds came out again and flew swiftly away. In four or five

---

minutes she returned, and after eyeing me suspiciously a short time flew again to the tree and, vanishing from sight in the hole, remained there. I was intently watching that small black spot in the bark to see her emerge, when a little boy came slowly sauntering past my bench, and glancing at him I found that his shrewd brown eyes were watching my face and that he had a knowing half-smile on his lips.

"Hullo, my boy!" I said. "I can see plainly enough what is in *your* mind. You know I'm watching a hole in the tree where a jackdaw has just gone in, and your intention is, when no one is about, to swarm up the tree and get the young birds."

"Oh, no," he returned. "I'm not going to climb the tree and don't want any young jackdaws. I always come to look because the birds breed in that hole every year. Two years ago I had a bird from the nest, but I don't want another."

Then at my invitation he sat down to tell me about it. One morning when he came the young had just come off, and he found one squatting on the ground under the trees, looking stupefied. No doubt when it flew out it had struck against

---

a trunk or branch and come down bruised and stunned.

He wrapped it up in a handkerchief and took it home to Deal and put it in a box; then mother got some flannel and made a sort of bed for it, and warmed some milk and they opened its beak and fed it with a teaspoon. Next day it was all right and opened its beak to be fed whenever they came near it, and in two or three days it began flying about the room and perching on their shoulders. Then he brought it back to Walmer and let it go and saw it fly off into the trees, but when he got home mother scolded him for having let it go when its parents were not about; she said it would die of starvation, and was going on at him when in flew the jackdaw and came flop on her shoulder! After that mother and father said they'd keep the daw a little longer, and then he could let it go at a distance where there were other daws about. By and by they said they'd let it stay where it was. Father liked a bloater for his tea, and there was nothing the jackdaw was fonder of, so he was always on the table at tea-time, eating out of father's plate. Then he got to be troublesome. He was always

---

watching for a door or window of the parlour to be opened to let the air in, and that was the room mother was so careful about, and every time he got in he'd fly straight to the mantelpiece, which was covered with photographs and ornaments. They were mostly those little things--pigs and dogs and parrots and all sorts of animals made of glass and china, and the jackdaw would begin to pick them up and throw them down on to the fender, and of course he broke a lot of them. That made mother mad, and she scolded him and told him to get rid of the bird. So he wrapped it up so as it shouldn't know where it was going and went off two or three miles along the coast, and let it go where there were other daws. It flew off and joined them, and he came home. That afternoon Jackie came back, and they wondered how he had found his way. Father said 'twas plain enough, that the bird had just followed the coast till he got back to Deal, and there he was at home. He said the only way to lose it was to take it somewhere away from the sea; so he wrapped it up again and took it to his Aunt Ellen's at Northbourne, about five miles from Deal. His aunt told him to carry

---

it to the park, where he'd find other daws and settle down. And that's what he did, but Jackie came back to Deal again that same day; the strangest thing was that mother and father made a great fuss over it and fed it just as if they were glad to have it back. Next day it got into the parlour and broke some more things, and mother scolded him for not getting rid of the bird, and father said he knew how it could be done. One of his pals was going to Dover, and he would ask him to take the bird and let it go up by the castle where it would mix with the jackdaws there, and that would be too far away for it to come back. But it did come back, and after that he sent it to Ashford, and then to Canterbury, and I don't know how many other places, but it always came back, and they always seemed very glad to see it back. All the same, mother was always scolding him about the bird and complaining to father about the damage it did in the house. Then one day Aunt Ellen came to see mother, and told her the best way to get rid of the daw would be to send it abroad; she said her husband's cousin, Mr. Sturge, was going out to his relations in Canada to work on their farm, and she would get

---

her husband to ask him to take the jackdaw. It would never come back from such a distant place. A week afterwards Mr. Sturge sent word that he would take the bird, as he thought his relations would like to have a real old English jackdaw to remind them of home. So one day Aunt Ellen came and took Jackie away in a small covered basket. The funniest thing was the way father went on when he came home to tea. "A bloater with a soft roe," he says; "just what Jackie likes! Where's the bird got to? Come to your tea, Jackie!"

"He's gone," says mother, "gone to Canada, and a good riddance, too!"

"Oh, gone, has he?" says father. "Then we're a happy family and going to lead a quiet life. No more screams and tears over broken chiny dolls! And if ever Billy brings another jackdaw into the house we'll dust his coat for him."

Here Billy interposed to say that if he ever made such a mistake again they could thrash him as much as they liked.

"Oh, yes," said father, "we'll thrash you fast enough; mother'll do it for the sake of her chiny toys and dolls."

BIRDS IN CORNISH VILLAGE 323

That put mother up. "You're in a nasty temper," she says, "but you know I miss the bird as much as you do!"

"Then," said father, "why the devil didn't you tell that sister of yours to mind her own business when she came interfering about my jackdaw! And that Sturge, he'll soon get tired of the bird and give it away for a pint of beer before he gets to Liverpool."

"So much the better," says mother. "If Jackie can get free before they take him aboard you may be sure he'll find his way back to Deal."

And that's what they went on hoping for days and days; but Jackie never came back, so I s'pose Mr. Sturge took him out all right and that

he's in Canada now.

THE END